FORECAST FOR JAPAN:
SECURITY IN THE 1970'S

FORECAST FOR JAPAN: SECURITY IN THE 1970'S

EDITED BY

James William Morley

CONTRIBUTORS

Donald C. Hellmann
Frank C. Langdon
James William Morley
Nathaniel B. Thayer
Martin E. Weinstein
Kenneth T. Young

PRINCETON UNIVERSITY PRESS
PRINCETON, NEW JERSEY

ISBN: 0-691-03091-X
LCC: 71-37578

This book was composed in Times Roman
Printed in the United States of America
by Princeton University Press, Princeton, New Jersey

Table of Contents

FORECAST FOR JAPAN:
SECURITY IN THE 1970'S

Introduction

For more than twenty years little has been said publicly about Japan's security policy. Japan was too poor and too timid to threaten anyone else and too well protected by the United States to feel itself threatened.

But Japan is no longer poor. It is now producing the third largest gross national product in the world. While it will not soon match the Soviet Union or the United States, it is already producing as much as all the rest of Asia combined and may well double its output again within the next five years. Nor is Japan any longer timid. Affluence has bred confidence, with the result that for the first time in a generation Japan is searching for an expanded international role. Coming at a time when the United States is withdrawing its own forces from the front lines in Asia, these developments pose questions not only for the Japanese, but for Americans as well. If Japan in 1975 produces a GNP more than twice as large as all the rest of Asia combined, as now seems likely, will it seek to build armed forces to match? What can Japan afford to do? What will it want to do?

Incontrovertible answers to such questions, of course, cannot be given. Nevertheless, it is important to seek as reliable forecasts as possible if the United States and other countries are to plan their own policies realistically.

Many matters may be thought to have a bearing on the path Japan takes. In this book, three in particular are examined: the fiscal and strategic attitudes of officialdom, the balance of domestic political forces, and pressures in the international environment. To secure as many independent points of view as possible, the views of six academic specialists are presented. Although an earlier version was prepared for the United States Arms Control and Disarmament Agency as part of a project organized by Howard Wriggins, John M. H. Lindbeck, and Stanley J. Heginbotham to seek independent assessments of Japanese and Indian security policies, the views set forth here are those of the authors and do not necessarily reflect the views of the United States Arms Control and Disarmament Agency or any other agency of the United States government.

I begin by analyzing the fiscal aspects of these questions in Chapter I. Finding that the extraordinary growth in the Japanese economy makes fiscally possible a quantum leap in Japanese military spending, I argue that it would be possible for the government to double or triple the share of the national budget it allocates to defense without substantially slowing down the economy or inhibiting a substantial increase in welfare spending now demanded by the people. It could thus, if it chose, in a very few years match the military power of the great European states and, should it wish to exert itself more, could in the same period of time overtake the People's Republic of China.

Barring unforeseen circumstances, however, it seems to me highly improbable that Japan will opt for such a stepped-up rearmament program within the next five years. The most likely projection of political, bureaucratic, and international pressures, I believe, will incline the government to hold the defense share of the budget steady, that is, at a level of somewhere around one percent of GNP. This would permit Japan by 1975 to be pretty much able to defend itself against conventional attack "by its own hands" as the premier has asked, but it would not give it the capacity to assert its power overseas and would not free it from dependence on the United States for advanced weapons supply and for nuclear deterrence. At this pace, it could not be a major military power before 1980.

In Chapter II Martin Weinstein examines these questions from the point of view of Japanese strategic thinking. He finds the origins of Japan's postwar security policy in the decisions taken by Prime Minister Yoshida and Foreign Minister Ashida in 1947. In the course of tracing the subsequent build-up of the Self-Defense Forces and the changes in the Japanese-American security relationship over the years, he notes a remarkable consistency in Japanese security thinking: that in a bi-polar world dominated by superpowers, Japan's best defense lies, not in going it alone with its own arms, but in relying on the United States for ultimate protection and concentrating its own energies on its economy. For the years immediately ahead, Weinstein sees no evidence of a weakening of this view.

Changes in the political leadership or in the political forces supporting this leadership could, of course, upset these projections. With this in mind, Nathaniel Thayer analyzes in Chapter III the structure and dynamics of Japanese domestic politics. Japan, he argues, is governed by a triumvirate of groups: big businessmen, top bureaucrats, and conservative party leaders. This triumvirate is largely insulated from mass public opinion and is supported by the postwar constitution and electoral system and by the balance of organized interest groups. Following an examination of the potential threats to this order, Thayer concludes that there is very little likelihood within the next five years that there will be any major changes in the institutional base of the system, the relative strength of the groups which participate in it, or the relative strength of the parties. The ruling Liberal Democratic party, he argues, should be able to continue in power.

Frank Langdon pushes the argument one step further, taking up in Chapter IV the most influential of the groups supporting the LDP: big business. What security policy will businessmen want over the next five years? He notes a growing desire for a more self-reliant security posture, more commensurate with Japan's real and growing economic capacities. This leads him to look particularly hard at those groups which are encouraging an expansion of the defense production effort: the Weapons Industry Association, Keidanren's Defense Production Committee, and the Defense Discussion Council. He finds, however, that these arms-makers have had little effect on the government's security policy in the past and are not likely to do so in the five years ahead. Arms production will increase, but not extraordinarily so, and the export of arms will not be significant, he believes, before 1975. Business will want no drastic change in security policy that would upset its peaceful pursuit of markets and sources of materials.

If the domestic political situation is likely to support a continuation of present policy, what of the external environment? Donald Hellmann copes with this question in Chapter V. He concludes that many of the foreseeable changes abroad will force Japan to begin to confront the problems of *realpolitik* which the American alliance has so far enabled it to avoid. The most serious of these problems, Hellmann believes, relates to China, for Japan is, he argues, locked willy-nilly into a long-term rivalry with China for leadership in Asia. On the other hand, the Sino-Japanese balance of power is not deemed likely to be upset within the next five years, and, unless that balance is upset, he concludes, Japan is not likely to shift its security posture.

Although the major contention of this book—that Japan is not likely seriously to change its security policy for at least the next five years—is shared by all the authors, assessments on many points differ. Hellmann differs with Weinstein's view that the Japanese government's "balanced defense" policies have been calculated responses to the realities of world politics, arguing instead that these conceptions have simply involved extension of past, Occupation-rooted policies by a leadership incapable of bold decisions operating within an international situation which left little effective room for choice.

In the face of an altered external environment, these policies will eventually have to go.

Kenneth Young is not so sure. In Chapter VI he presents his analysis of Japan's Southeast Asian involvement, agreeing that Japan is increasingly confronted with *realpolitik* there; but he does not believe that the relationship with China needs to be a confrontationist one and concludes that these countries are not likely to welcome a Japanese military interest. Instead, what they will desire from Japan will be increasing social, economic, and political cooperation in nation-building at home and region-building abroad—a desire that should fit very well with the predilections of the Japanese themselves to seek security in the region through non-military means.

Drawing on these independent projections of bureaucratic thinking, internal political forces, and the external security environment, I offer in the final chapter a primary forecast of the security policy that I believe Japan is most likely to follow over the early part of this decade. This is a personal rather than a collective view that Japan will be neither militarist nor pacifist, neither chauvinist nor isolationist. Instead, it will follow a moderate, constructive policy of "balanced defense," continuing to keep its military build-up and security commitments rationally proportional to its national priorities for political stability, economic growth, social welfare, and international conciliation.

Given Japan's enormous economy and intellectual vigor, this is fortunate indeed, for it means that both as a model and as an influence Japan can play an extremely effective role in stabilizing the peace of Asia. At the same time, it must be recognized that there are pressures at home and abroad that could push Japan onto a more militarily assertive course and that these pressures may well grow as the decade wears on.

For the United States this forecast has important implications, for it suggests that if the Japanese people are to continue to find satisfaction in their present security policy, we should give them our fullest understanding and cooperation. I suggest also that we should seize this extraordinarily favorable opportunity to work out cooperatively with our most important ally in Asia a broad policy of detente for the entire Asian and Pacific region.

J.W.M.

October 1971

CHAPTER I

Economism and Balanced Defense

JAMES WILLIAM MORLEY

WITH the rise of Japan to the position of third-ranking economy in the world and the germination of a new nationalism in the land, there is a growing expectation in many quarters that Japan will soon want to "rearm" rather sharply and to reach out militarily for the defense of its expanding interests around the world, particularly in Asia.

There are those on the Right who look forward to such a course, hoping that Japan will contribute a new force for the containment of communism or at least "share" some of the "burdens" that the United States wishes to lay down. There are others on the Left who are apprehensive of such a course. The Chinese have been among the most extreme exponents of this view, Premier Chou En-lai reportedly arguing, for example, that under American prodding, Japan is planning greatly to expand its armed forces with the intent of either fighting America's battles in Asia or reviving its own imperial sphere.[1] In between are a growing number of realistic commentators who feel unsure how far Japan will go, but do not believe, whether one approves or disapproves, that any state as wealthy, dynamic, and ambitious as Japan is today will long be content to bow its head. Is it not the common experience of mankind, they ask, for nations that acquire wealth to seek also power? Is it not likely, therefore, as the Japanese expand their economy at home and acquire ever greater interest in trade and investment abroad, that they will be less willing to rely on the United States for their ultimate security, more willing to build up their own armed forces, and more interest-

[1] See, for example, Ross Terrill, "The 800,000,000: China and the World, Part II," *The Atlantic Monthly*, January 1972, pp. 40-41.

ed in deploying these forces, selling their arms and pursuing the alliance politics that characterize the behavior of all of Japan's great neighbors in Asia today? Is it not almost inevitable, to borrow the words of Herman Kahn, that Japan will become "nuclear power number 6?"[2]

The significance of such premonitions can perhaps best be appreciated in comparative economic terms. Of course, one cannot equate levels of spending directly with military power, which must include geographical, political, psychological, and other assessments in addition to the composition of forces for which the money is spent. But defense budgets may serve as a very rough guide to capacity and intent. According to data available for 1967, the nations of the world fell into three categories of defense spending: the superpowers, including only the United States and the Soviet Union, which spent annually more than $50 billion each, far more than all the other countries put together; the major powers, including the People's Republic of China, France, West Germany, and Britain, which spent roughly $5-$7 billion each; and the minor powers, including all the rest, which spent less than $2 billion each.[3]

Japan, of course, was in the minor power class, spending only slightly more than $1 billion (see Table 1); but in view of the size of its economy and the rapidity of its growth, all it would need to do to become a military superpower by 1975 would be for it to allocate about 10 percent of its GNP annually for military purposes. This can be demonstrated from the data presented in Tables 2 and 3. In 1968 Japan's GNP reached $141.9 billion. This was the result of a fairly constant

[2] Emerson Chapin, ed., *Japan and the United States in the 1970's* (New York: Japan Society, 1971), a full record of discussions at Wingspread, Racine, Wis., December 2, 1969, led by James C. Abegglen, Herman Kahn, and James W. Morley, p. 28.

[3] From data provided by the Institute for Strategic Studies in *The Military Balance, 1968-1969* (London: Institute for Strategic Studies, 1968), p. 57.

TABLE 1

DEFENSE EXPENDITURE AS SHARE OF GNP AND SHARE OF NATIONAL BUDGET, 1960-1969

	DEFENSE RELATED EXPENDITURES		
Year	*In Bil. Dollars*[1]	*As % of GNP*[2]	*As % of Budget*[3]
1960	0.4	1.00	
1961	0.5	0.95	
1962	0.6	1.01	8.50
1963	0.7	1.00	8.05
1964	0.8	0.98	8.49
1965	0.8	0.97	8.24
1966	1.0	0.94	7.74
1967	1.1	0.90	7.47
1968	1.2	0.83	7.28
1969	1.3	0.84	7.22
Average annual change		–0.02	–1.17
Average annual share		0.9	7.87

[1] Data for 1962-1967 from Table 7; for 1968 from Japan, Ministry of Finance, Bureau of the Budget, *Zaisei tokei, Showa 43 nendo* (Public Finance Statistics: 1968), pp. 287-90; and for 1969, *Zaisei tokei, Showa 44 nendo* (Public Finance Statistics: 1969), pp. 255-56; using a conversion rate of Y360 = $1.00, the rate then obtaining.

[2] Calculated using GNP data in Table 2.

[3] Data for 1962-1967 from Table 7; for 1968 and 1969 calculated from sources listed in footnote 1 above.

and unprecedentedly high rate of growth augmented by inflation over many years—an average of 16.1 percent annually over the past decade. If the Japanese economy continues to grow at this rate, by 1975 the GNP will climb to more than $470 billion. If 10 percent of this were allocated to military purposes, these expenditures by 1975 could reach $47 billion annually, making Japan the third largest military spender in the world. Clearly, such appropriations could give Japan a nuclear-armed firepower perhaps five or more times greater than that of any other of the major powers, including mainland China, and within range of those of the USA and the USSR.

TABLE 2

GROWTH IN GNP, 1959-1968

CALENDAR YEAR	GNP AT CONSTANT PRICES[1]			GNP AT CURRENT PRICES			WORLD RANK OF GNP AT CURRENT PRICES
	Billions of Yen	*Billions of Dollars*[2]	*% Change*	*Billions of Yen*	*Billions of Dollars*[2]	*% Change*	
1959	17,257.7	47.9	9.2	12,926.3	35.9	12.2	6
1960	19,698.7	54.7	14.1	15,499.2	43.1	19.9	6
1961	22,765.9	63.2	15.6	19,125.5	53.1	23.4	6
1962	24,228.1	67.3	6.4	21,199.2	58.9	10.8	6
1963	26,785.3	74.4	10.6	24,464.0	68.0	15.4	6
1964	30,361.2	84.3	13.3	28,832.9	80.1	17.9	6
1965	31,713.9	88.1	4.5	31,792.9	88.3	10.2	6
1966	34,899.9	96.9	10.0	36,557.4	101.5	15.0	6
1967	39,403.1	109.5	12.9	43,038.9	119.6	17.7	4
1968	45,096.1	125.3	14.4	51,092.0	141.9	18.7	3
Average for ten years, 1959-1968:			11.1			16.1	

Source: Japanese Government, Economic Planning Agency
[1] GNP at constant 1965 prices
[2] Converted at Y360.00 = $1.00, the rate then obtaining.

TABLE 3

A PRIMARY FORECAST OF GROWTH IN GNP, 1969-1975

YEAR	AVERAGE ANNUAL % CHANGE OF GNP AT CONSTANT PRICES, 1959-1968[1]	GNP AT CONSTANT PRICES[2]		AVERAGE ANNUAL % CHANGE OF GNP AT CURRENT PRICES, 1959-1968[1]	GNP AT CURRENT PRICES	
		Billions of Yen	*Billions of Dollars*[3]		*Billions of Yen*	*Billions of Dollars*[3]
1969	11.1	50,101.8	139.2	16.1	59,317.8	164.8
1970	11.1	55,693.1	154.7	16.1	68,868.0	191.3
1971	11.1	61,875.0	171.9	16.1	79,955.7	222.1
1972	11.1	68,743.1	223.2	16.1	92,828.6	301.4
1973	11.1	76,373.6	248.0	16.1	107,774.0	349.9
1974	11.1	84,851.1	275.5	16.1	125,125.6	406.2
1975	11.1	94,269.6	306.1	16.1	145,270.8	471.7

[1] From Table 2 above.

[2] GNP at constant 1965 prices.

[3] For 1969-1971, converted at Y360.00 = $1.00, the rate then obtaining; for 1972-1975, converted at Y308.00 = $1.00, the revised rate set on December 20, 1971.

This is a shocking possibility, to be sure, but not on the face of it absurd if one considers the common experience of mankind that as nations grow richer, they generally grow prouder, acquire more international interests, and seek greater military power to defend them; and if one also remembers that the share of GNP which each of Japan's great neighbors is allocating for defense is roughly of this dimension: 7 percent for the People's Republic of China, 10 percent for the United States, and 14 percent for the USSR.[4] The superpower option in today's world would appear, in fact, to be the option of conventional common sense.

To become a major military power by the mid-1970's would be, of course, far easier. It would require the allocation of only 2 percent of the GNP for this purpose. On the same assumption of continued GNP growth as before, a 2 percent allocation for defense in 1975 would yield $9.4 billion. This would compare very favorably with the $5-$7 billion annual expenditures of the other major military powers in recent years. (See Tables 2 to 5 for GNP growths.)

Small wonder, then, that for some time many observers have been expecting Japan to adopt one or another version of these accelerated rearmament policies. The paradox is that it has not done so. Instead, for more than a decade the annual share of GNP devoted to military purposes has been, not 10 percent nor even 2 percent, but an average of 0.9 percent, and even that share has been steadily declining (at the rate of 0.02 percent a year). In 1967 this amounted to $1.1 billion, only 1.5 percent of the U.S. allocation, 15 percent of that of mainland China, and less than those of such small countries as Czechoslovakia or Australia.[5] Since 1967 Japan

[4] These estimates are derived from the data for 1967-1969 presented in *ibid., passim*; and for the PRC from *Kokubo* (National Defense), August 1969, p. 289, and Table 2 in Chapter V below.

[5] *Kokubo,* August 1969, p. 289.

has been working to complete its Third Defense Build-up Plan. By the end of 1971 it was hoped to provide a Ground Self-Defense Force of approximately 180,000 uniformed personnel, organized into thirteen divisions and equipped with medium tanks, artillery, and Hawk surface-to-air missiles; a Maritime Self-Defense Force of approximately 142,000 tons, its largest units being destroyers, submarines, and anti-submarine aircraft; and an Air Self-Defense Force of approximately 880 aircraft, centering on F104J and F4 Phantom interceptors, NIKE-AJAX surface-to-air missiles, and the BADGE Automatic Radar (warning and control) system. This force, not yet fully achieved, has the limited capability of simply enabling Japan to hold its own home islands against an attack by conventional arms for a month or so while it waits for external support from the United States or the United Nations. It is not in any sense designed to play an independent role in the international balance of power and not even to protect Japan against advanced weaponry or a prolonged assault.

The relative modesty of this effort should not be taken to imply that the Japanese have felt no need at all to protect themselves against the dangers of an assault by a great power or a prolonged war. Obviously they have thought it better to meet these dangers by pursuing a conciliatory foreign policy and relying on the security treaty with the United States rather than to try to build the vast war machine necessary to confront the world alone. Nor does it mean that the Japanese have felt no need at all to shake their dependence on American military power or to strengthen their policy abroad with their own hands.

What it does mean is that the Japanese security policy has been based on a uniquely broad and economically centered conception that can best be described as one of balanced defense. As the conservatives who rule Japan see it, the security

TABLE 4

GROWTH IN GNP PER CAPITA, 1959-1968

CALENDAR YEAR	POPULATION[1] *(1,000)*	WORLD RANK OF GNP PER CAPITA[2]	GNP PER CAPITA AT CURRENT PRICES[3] *Yen (1,000)*	*Dollars*
1959	92,640		140	388
1960	93,420	21	166	461
1961	94,290		203	563
1962	95,180		223	619
1963	96,160		254	707
1964	97,180		297	824
1965	98,270	21	324	899
1966	99,050		369	1,025
1967	100,240		429	1,193
1968	101,408		504	1,399
Average Annual Increase:			15.3%	

[1] Office of the Prime Minister, Bureau of Statistics, *Monthly Statistics of Japan*, No. 93, March 1969, p. 3.

[2] Japan Research Center estimates in *Nihon keizai*, July 15, 1969.

[3] Derived by dividing GNP at current prices in billions of yen and dollars for each year in Table 2 above by population estimate for corresponding year.

of a country depends not so much on forces in being—unless the threat is immediate—as on that country's total strength, viewed as a compound of the moral fiber of the people, the productivity of the economy, and the stability of the polity. It is to an overevaluation of the importance of forces in being and an underappreciation of the importance of total strength that they attribute the defeat of Japan in the Pacific war. They do not intend to make the same mistake again. Consequently, rather than rush headlong into a vast rearmament program, they have adopted a unique policy in which the allocations for security needs are kept commensurate not so much with the practices of their great neighbors as with a realistic estimate of the threats in Japan's security environment, the strength of Japan's other needs, and above all the

TABLE 5

A PRIMARY FORECAST OF GROWTH IN GNP PER CAPITA, 1969-1975

CALENDAR YEAR	POPULATION[1] (1,000)	GNP PER CAPITA AT CURRENT PRICES[2]		WORLD RANK OF GNP PER CAPITA[3]	COMPARISON WITH U.S.	
		Yen (1,000)	*Dollars*		*GNP Per Capita at Current Prices*	*Year*
1969	102,569	578	1,607		$1,605[4]	1947
1970	103,744	664	1,844	17		
1971	104,929	762	2,117			
1972	106,140	874	2,840			
1973	107,372	1,004	3,260			
1974	108,635	1,152	3,739		3,520[4]	1965
1975	109,925	1,322	4,291	9	6,209[5]	1975

[1] Ministry of Health and Welfare, Institute of Population Problems.

[2] Derived by dividing GNP at current prices in billions of yen and dollars for each year in Table 3 above by population estimate for corresponding year.

[3] Japan Research Center estimates in *Nihon keizai*, July 15, 1969.

[4] In dollars at current prices, derived by dividing population estimate by GNP at current prices in billions of dollars, using data in U.S., Congress, Joint Economic Committee, *1967 Supplement to Economic Indicators* (Washington, D.C.: GPO, 1967), pp. 7, 15.

[5] Japan Research Center estimate in *Nihon keizai*, July 15, 1969.

requirements of its overall policy of economism which emphasizes economic growth.

The result has been a public fiscal formula to which they have adhered firmly for more than a decade. One element of the formula has been to emphasize the private sector by keeping the national budget small. For the past five years that budget has been held to an average of only about 12 percent of GNP, the average annual change being no more than plus or minus 0.6 percent (see Table 6). Another element has been to allocate the public moneys within the budget into eleven shares, each of which also has been held remarkably steady, gaining or losing over the five-year period from 1962 to 1967, for example, an average of no more than 1.74 percent of the total (see Table 7).

The share which varied the most, that for industrial development, changed only by 4.33 percent of the budget. If one

TABLE 6

NATIONAL BUDGET EXPENSES AS PERCENTAGE OF GNP

	NBE (A)[1] (*mil.-yen*)	*GNP (B)*[2] (*mil.-yen*)	*A/B* (*percent*)	*Government Consumption*[3] (*percent of GNP*)	*Government Investment*[4] (*percent of GNP*)
FY 1962	2,556.6	21,199.2	12.1	8.7	3.4
1963	3,044.3	24,464.0	12.4	9.2	3.2
1964	3,311.0	28,832.9	11.5	9.0	2.5
1965	3,732.0	31,792.9	11.4	9.2	2.2
1966	4,477.1	36,557.4	12.2	9.3	2.9
1967	4,950.9	43,038.9	11.5	9.0	2.5
Average annual ratio			11.9	9.1	2.8
Average annual change			±0.6	±0.3	±0.5

[1] Data for 1962 and 1963 from Office of the Prime Minister, Bureau of Statistics, *Japan Statistical Yearbook, 1965* (1966), pp. 476-81; and for 1964-1967 from *JSY, 1967* (1968), pp. 472-74.

[2] See Table 2 above.

[3] Office of the Prime Minister, Bureau of Statistics, *Monthly Statistics of Japan*, No. 93, March 1969, p. 117.

[4] Calculated as the difference between column 4 and column 3.

TABLE 7

NATIONAL BUDGET EXPENDITURES, 1962-1967

	FY 1962 (closed) % total	*FY 1963 (closed)* % total	*FY 1964 (closed)* % total	*FY 1965 (closed)* % total	*FY 1966 (budget)* % total	*FY 1967 (original budget)* % total	*FY 1962-1967* Change in %
TOTAL	100.00	100.00	100.00	100.00	100.00	100.00	
National administration	8.64	8.51	8.74	8.88	7.37	7.40	1.51
Local finance	19.13	19.28	19.32	19.34	18.70	18.65	0.69
National defense	8.50	8.05	8.49	8.24	7.74	7.73	0.77
Foreign obligations (incl. reparations)	1.14	0.79	0.77	0.53	0.70	0.68	0.61
Land conservation and development	18.37	17.68	18.45	19.19	18.73	18.92	1.51
Industrial development	7.39	7.14	8.05	8.27	11.47	9.13	4.33
Education & culture	11.95	12.15	12.35	12.67	12.16	12.23	.72
Social security, etc.	14.45	14.91	15.61	17.20	16.42	16.89	2.44
Pensions	4.82	4.38	4.55	4.22	3.96	4.04	.86
National debt	2.63	3.76	1.36	0.35	1.01	2.33	3.41
Other (incl. reserves)	2.99	3.36	2.31	1.11	1.67	2.01	2.25
					Average change in share:		1.74

Sources: For 1962 and 1963 percentages calculated from data in Office of the Prime Minister, Bureau of Statistics, *Japan Statistical Yearbook, 1965* (1966), pp. 476-81; and for 1964-1967 from *JSY, 1967* (1968), pp. 472-74.

adds up the total changes in shares each year among these purposes as given in Table 7, he finds that for the five-year period between 4.62 percent and 8.65 percent of the annual budget has been reallocated each year, the average being 6.24 percent. Assuming a budget stationary at 12 percent of GNP, this equals between 0.5 percent and 1.0 percent of GNP a year, with the average running about 0.7 percent.

After pensions and foreign obligations (including reparations), the share of the budget which has been held steadiest has been that for defense, its share of the annual budget changing on the average only about 0.33 percent and declining nearly every year from a high of 8.5 percent to a low of 7.7 percent of the total budget, or 1.01 percent and 0.90 percent of the GNP (see Table 1).

If Japan continues for the next five years pretty much as it has in the past five, there is no accelerated rearmament in prospect. The national budget may be expected to be held at about 12 percent of GNP, and the defense share can be expected to decline each year at the annual average rate of the past ten. This would mean for 1970 an allocation of 0.82 percent of the GNP or approximately $1.6 billion, declining presumably in regular annual stages to 0.72 percent of GNP in 1975, but increasing in that year in absolute terms concomitantly with the growth in GNP to $3.4 billion.

This may be a bit too precise. It could hardly be argued that a declining share is necessary to the policy of balanced defense. It is perhaps more reasonable to suppose that as military affairs assume greater importance, the Defense Agency will acquire a politically more powerful leadership and that, therefore, in the years ahead, the agency will be able to check the slippage and hold its share more or less constant, say roughly at the average level of the past decade, about 0.9 percent. Given our projections of the growth of GNP, this would yield something on the order of $4.2 billion in 1975, or a total over the period 1971-1975 of approximately $15.7 billion (see Table 8). Remarks of Yasuhiro Nakasone, when serving as director-general of the National Defense Agency, indicate that current government planning is indeed proceeding along these or very similar lines.[6]

In military terms such spending could hardly put Japan in the major power class by 1975, but it would advance Japan halfway toward it in what might be called a middle-power

TABLE 8

PRIMARY FORECAST OF DEFENSE-RELATED EXPENDITURES, 1971-1975
(Under the Traditional Policy of Balanced Defense, Assuming a Constant Share of GNP and a Constant Growth Rate)

		AT CURRENT PRICES		AT 1965 PRICES	
Year	*Defense Share of GNP*[1] *(%)*	*GNP Rate of Growth*[2] *(%)*	*Annual Defense*[3] *($ bil.)*	*GNP Rate of Growth*[4] *(%)*	*Annual Defense*[5] *($ bil.)*
1971	0.9	16.1	2.0	11.1	1.5
1972	0.9	16.1	2.7	11.1	2.0
1973	0.9	16.1	3.1	11.1	2.2
1974	0.9	16.1	3.7	11.1	2.5
1975	0.9	16.1	4.2	11.1	2.8
Totals			15.7	11.1	11.0

[1] Average annual share, 1960-1969, as calculated in Table 1.
[2] Average annual rate of growth of GNP, 1960-1969, at current prices as calculated in Table 2.
[3] Calculated at the rate of 0.9% of the primary forecast of growth in GNP at current prices at the rate of 16.1% a year as given in Table 3.
[4] Average annual rate of growth of GNP, 1960-1969, at 1965 prices as calculated in Table 2.
[5] Calculated at the rate of 0.9% of the primary forecast of growth in GNP at 1965 prices at the rate of 11.1% a year as given in Table 3.

[6] Selig S. Harrison reported in the *Washington Post*, June 23, 1970, that "Nakasone has been talking lately of $15.3 billion in defense spending between 1972 and 1976," presumably in connection with planning for the Fourth Defense Build-up Plan. This figure is close to, but slightly smaller than, one resulting from the above forecast if extended for these years using the exchange rate then obtaining ($15.9 billion), suggesting that Nakasone may have been projecting a stationary budget share as we have, but a slightly lower growth rate for the economy.

range that it would occupy all by itself, perhaps ranking number seven in the world, spending about half as much as the mainland Chinese and about half as much as the rest of Asia combined.

There are then three alternatives for defense policy that Japan's burgeoning economy places before it in the first half of the 1970's: a superpower option, requiring a defense share of about 10 percent of GNP; a major power option, requiring a 2 percent share; and a middle-power option, requiring only the maintenance of its present less than 1 percent share. Which will it choose?

In subsequent chapters the strength of continuity and the possibilities for change will be analyzed in terms of the security concepts prevalent in Japan, the political leadership that can be expected, the business pressures that are likely, and the effects that shifts in the international environment may occasion. Here let us concentrate on the probable economic and bureaucratic determinants. Among these, the most important would appear to be the relationship of military expenditures to economic growth.

That stable, high growth remains *the* national imperative can hardly be doubted. It is deeply ingrained in the national consensus. A century ago when Japan threw off its feudal institutions and set out on the path of modernization, it determined on two objectives: wealth and military power, assuming that both could be attained simultaneously. One of the great lessons of World War II for most Japanese was that military power on such a scale was not needed and, in any event, could not be built unless and until the country had fully developed its economy. Economic growth is seen as the ground that makes all things possible. This lesson of history has been confirmed by the experience of the past two decades when peace has been maintained without large standing forces and the growth rate has steadily been accelerated, rising at constant prices from an annual average of 8.6 percent in

1951-1955 to 9.1 percent in 1955-1960 to 9.7 percent in 1960-1965 and 13.1 percent in 1965-1970.

Throughout the 1950's and 1960's Japanese officials and academic economists alike were constantly fearful that such a high rate of growth could not be sustained. They were, therefore, adamantly opposed to any increase in the share of the GNP allocated to defense since this was seen as a diversion of resources from productive uses and consequently a serious threat to the growth that they felt was so necessary, but so shaky. Premier Ikeda, in 1960, shocked the nation with the idea that the government could devise plans to bring about a doubling of the income within ten years. The fact is, income was quadrupled. By now high growth, unusual though it is among advanced industrial nations, is assumed by Japanese experts to be on a sound basis. Before the international monetary crisis of 1971, the Economic Planning Agency forecast an average annual real growth rate in the years ahead of 10.6 percent, the Long Term Credit Bank of Japan said 11.3 percent, and the Japan Economic Research Center, 12.4 percent.[7] As a result of the monetary crisis, those rates were not achieved in fiscal 1971, but informed opinion expects rapid growth to be resumed in 1972. Our own estimate of 11.1 percent, therefore, seems not unreasonable.

These new conditions of economic confidence make it possible to examine more calmly than ever before the real relationship between the military share and the economic growth rate. The truth seems to be that, given the size and strength of the Japanese economy, within reasonable limits changes in the quantity of military spending may not produce wholly unacceptable effects on the growth rate. The analytical tool here is the gross incremental capital-output ratio, a figure that states the ratio between the gross domestic investment in a given year and the gross increment in GNP for the same

[7] Japan Economic Research Center, *Japan's Economy in 1975* (Tokyo, 1970).

TABLE 9

GROSS INCREMENTAL CAPITAL–OUTPUT RATIO, 1959-1968

Year	*GNP In bil. $ At current prices*[1] *(A)*	*Gross Domestic Investment In bil. $ At current prices*[2] *(B)*	*As % of GNP (B/A)*	*Gross Incremental GNP In bil. $ At current prices*[3] *(C)*	*Gross Incremental Capital-Output Ratio (B/C)*
1959	35.9	10.7	29.8		
1960	43.1	14.5	33.6	7.2	2.0
1961	53.1	21.5	40.5	10.0	2.2
1962	58.9	21.1	35.8	5.8	3.6
1963	68.0	24.3	35.7	9.1	2.7
1964	80.1	29.1	36.3	12.1	2.4
1965	88.3	29.3	33.2	8.2	3.6
1966	101.5	34.4	33.9	13.2	2.6
1967	119.6	44.9	37.5	18.1	2.5
1968	141.9	54.7	38.5	22.3	2.5

[1] See Table 2 above.
[2] Data for 1959-1968 from Office of the Prime Minister, Bureau of Statistics, *Monthly Statistics of Japan*, No. 96, June 1969, p. 117.
[3] Calculated from data in column (A).

year.[8] As indicated in Tables 9 and 10, the ratio for Japan during the period 1971-1975 will probably run about 2.6. This means that for every yen invested or disinvested in any one year during the period, the GNP can be expected to rise or fall about 0.4 yen; or, in terms of percentages, that for every 1 percent of GNP added to or subtracted from the gross domestic investment, the GNP increment can be expected to rise or fall about 0.4 percent.

Now, to take the extreme case, let us suppose that all new money for an expanded defense share would be taken away from investment, whether public or private, and that, as is standard accounting procedure, none of the additional expenditures resulting could reasonably be considered invest-

[8] For a discussion of the uses and limits of this ratio see Simon Kuznets, *Modern Economic Growth: Rate, Structure, and Spread* (New Haven: Yale University Press, 1966), pp. 258-62.

TABLE 10

A PRIMARY FORECAST OF THE GROSS INCREMENTAL CAPITAL–OUTPUT RATIO, 1969-1975

Year	*GNP In bil. $ At current prices*[1] *(A)*	*Gross Domestic Investment In bil. $ At current prices*[2] *(B)*	*As % of GNP (B/A)*	*Gross Incremental GNP In bil. $ At current prices*[3] *(C)*	*Gross Incremental Capital-Output Ratio (B/C)*
1969	164.8	58.3	35.4	22.9	2.5
1970	191.3	67.7	35.4	26.5	2.6
1971	222.1	78.6	35.4	30.8	2.6
1972	301.4	106.7	35.4	79.3	1.3
1973	349.9	123.9	35.4	48.5	2.6
1974	406.2	143.8	35.4	56.3	2.6
1975	471.1	167.0	35.4	65.5	2.6

[1] See Table 3 above.

[2] Data for 1969-1975 are projected on the assumption that the percentage of GNP devoted to gross domestic investment each year will continue to be the average annual percentage devoted to that purpose in 1959-1968 or 35.4%. (An alternative projection might be made on the assumption that this percentage will continue to increase at the same average annual rate as in 1959-1968 or 0.97% annually, but it is thought unlikely that such a growth will be sustained.)

[3] Calculated from data in column (A).

ment—assumptions that are demonstrably false, but other assumptions would only strengthen our argument. We find then that the most serious effect that an increase in defense allocations could have on the economy would be a 0.4 percent decline in GNP growth for every 1 percent of GNP added to the annual defense expenditures. In short, a decision to become a major military power by 1975 by raising the defense share of GNP roughly one additional percent might be expected at most to depress the potential growth rate in that year by only 0.4 percent, that is, to lower it from 11.1 percent to 10.8 percent in real terms. And a decision to seek superpower spending status by that year—allocating, say, 10 percent of GNP or nine percent in addition to what is currently allocated—could be expected to depress the potential growth rate in that year by only 3.6 percent or to lower it in real terms from 11.1 percent to 8.6 percent.

These are still handsome growth rates and indicate that in an emergency the Japanese economy can sustain an extraordinarily large and rapid increase in defense spending without seriously crippling its productivity. To put it another way, Japan can do economically what it militarily wants to if it wants to badly enough. But as long as sustained high economic growth is the central objective of Japanese governmental policy, the potentially depressing effect of accelerated defense spending on that growth can be expected to serve as a strong deterrent to taking any great military leap forward.

Moreover, even if a lower growth rate were tolerable, resulting in a smaller economic pie than desired, to cut a large slice of that pie for defense would inevitably reduce the size of the slices available for other purposes. How acceptable would this be?

There are after all only three sources from which a larger budget or GNP share for defense can be derived: government borrowing, additional taxation, or the reallocation of national budget shares. None of these would be popular. A resort to any one of them would require a major decision and one certain to generate varying degrees of opposition under present circumstances.

Presumably, borrowing would be the easiest. The total Japanese government debt is relatively small, its servicing in the period 1962-1967, for example, requiring no more than 4 percent of the national budget in any one year. Nor are current government bond issues large, the percentage of national revenue provided by bonds falling from 16.9 percent in 1966 to 7.2 percent in 1969.[9] These figures suggest that, should it be felt necessary, Japan could finance a more rapid rearmament by borrowing, offsetting any possible inflationary effects by a variety of open market or other operations.

[9] Japan, Ministry of Finance, Budget Bureau, *The Budget in Brief: Japan, 1969* (n.d.), p. 46.

Seeking to raise a large additional amount by taxation would be more unpopular. This is not because the tax level is already excessively high compared with other advanced countries, but rather because of the relative stability of the gross incremental government expenditure-tax ratio, to which the public has become accustomed. The fact that Japan has been raising a good proportion of its revenues by progressive income and corporation taxes in a rapidly growing economy has meant that, as incomes rise, the tax return also tends to rise even though the terms of the levy are held constant; in fact, the tax return increases faster than the GNP. Since Japan has at the same time been pursuing a restrictive budget policy, if the tax laws were held constant, the tax returns would annually exceed those required for national expenditures. Consequently, it has become possible, indeed customary, to lower the taxes annually, thus keeping the proportion of GNP taken up by national tax for 1957-1966, for example, at an annual average of about 11.0 percent and varying the average from year to year no more than plus or minus 0.4 percent (see Table 11). For every additional 1 percent of GNP that might be spent on defense (or any other purpose), the tax revenues—if they were the sole source of financing—would have to be increased by 1/11 or approximately 9.9 percent. Changes of this scale would not be unprecedented. They occurred in 1958-1960, for example, but they would reverse a steady trend since 1960 of annual tax reduction and presumably would require a strong government to carry them off in the face of easily foreseen opposition.

The feasibility of reallocating the national budget shares depends in a political sense on the strength of the constituencies each of the shares represents. Several of the traditional constituencies would seem to be facing the prospect of declining budget shares. These include the farmers for their rice price supports and the industrialists for various favors to big

TABLE 11

NATIONAL TAX REVENUE AS PERCENT OF GNP, 1957-1966

Year	*GNP (bil. yen)*[1]	*National Tax Revenue (bil. yen)*[2]	*Tax-GNP Ratio (%)*
1957	11,020.8	1,201.8	10.9
1958	11,341.6	1,190.8	10.5
1959	12,926.3	1,372.4	10.6
1960	15,499.2	1,801.5	11.6
1961	19,125.5	2,227.7	11.7
1962	21,199.2	2,390.7	11.3
1963	24,464.0	2,731.7	11.2
1964	28,832.9	3,159.2	11.0
1965	31,792.9	3,279.7	10.3
1966	36,557.4	3,663.0	10.0
Average annual ratio			10.9%
Average change in annual ratio			± 0.4%

[1] For 1957 and 1958, Japan, Office of the Prime Minister, Bureau of Statistics, *Monthly Statistics of Japan*, No. 93, March 1969, p. 117; for 1959-1966, see Table 2 above.

[2] Japan, Ministry of Finance, Bureau of the Budget, *Zaisei tokei*: Showa 43 nendo (Public Finance Statistics: 1968), p. 310.

business. But a number of others are pressing for larger expenditures, and, it may be assumed, for larger shares. These include the underprivileged, demanding increased social security; communities all over the country, calling for more and more public works, particularly roads, ports and harbors, and housing; educational circles, business, and the public at large, demanding more support for higher education, which is only now recovering from the upheaval of 1967-1969, and for research and development, which has been hitherto slighted; urbanites, incensed by public pollution and wanting the government to do something about it; and various foreign governments, who are pressuring Japan for increased aid.[10]

[10] For the dimensions of some of these demands, see James W. Morley, "Growth for What? The Issue of the Seventies," in Gerald L. Cur-

In few of these areas have the constituencies been able to secure any kind of firm commitments to larger budget shares from their bureaucratic sponsors. One exception may be the foreign governments, who have persuaded the Japanese government to commit itself to increase its foreign aid from 0.7 percent of GNP in 1969 to 1 percent of GNP by 1975.[11] However, what effect this will have on the aid share of the national budget remains to be seen, since the cabinet ministers have refused to promise to raise the governmental portion of this aid to 0.7 percent of GNP as requested. As a conclusion, perhaps one can only go so far as to forecast that for the roughly 6.2 percent of the budget or 0.7 percent of the GNP which is on the average annually reallocated among the budget shares, there will continue to be many competitors.

Each of these competitors, of course, has its own bureaucratic champion, the Japanese government being administered in a very real sense by bureaucracies rather than a bureaucracy, with each of the separate bureaus being internally hierarchical and externally competitive with all other bureaus. This is conducive to following precedent. In Japan as in the United States, for example, the ground from which each year's national budget is constructed is the budget of the previous year, each bureau customarily being advised by the Ministry of Finance to propose an allocation within so many percentage points of the previous year's allocation. These proposals, which are inevitably maximal, are then cut and sanded until they fit as neatly as possible into the overall figure which the ministry and finally the cabinet and ruling party have concluded the anticipated growth rate makes possible: that is, into a figure that is about 12 percent of the antici-

---

tis, ed., *Japanese-American Relations in the 1970s* (Washington, D.C.: Columbia Books, for the American Assembly, 1970), pp. 48-93.

[11] Cabinet decision of May 12, 1970, as reported in *Japan Report,* Vol. 16, No. 12, June 16, 1970, pp. 1-2.

pated GNP. In this process, Japanese officials are no more immune than American officials to the Hitch and McKean principle that "there is often severe personal penalty for originating mistakes, yet little or no penalty for perpetuating past decisions."[12]

The bureaucratic structure of the Japanese government is also conducive to stalemate. For a broad policy change, such as a significant shift in the defense share of the budget, to be initiated successfully by the bureaucracy would require a change in bureaucratic thinking in which a number of bureaus formerly competitive would subordinate their parochial interests to an overriding concern for defense. Judging from the experience of the 1930's, this might come about if there were to form within a number of the bureaus groups of defense-minded bureaucrats who by informal liaison could achieve preponderant influence at a number of strategic points, including the Defense Agency, the Foreign Ministry, the Finance Ministry, and the planning agencies. No such factional movement is now apparent and it is unlikely to develop without strong supporting activity in the political parties and the public at large. Conceivably, it might also come about by a broad generational change in bureaucratic leadership. By the early 1970's the preponderant part of officialdom will consist of younger men who did not fight in the Pacific war and who are less saddled with the confusions of defeat. They may indeed be increasingly prepared to hearken to a call to arms. But the generation who will hold the posts of section chief, bureau chief, and above in the early 1970's will continue to be the generation that did fight in the war. Its disillusionment with war is deep. It is not likely to want to sacrifice too much to increase Japan's military power.

This analysis indicates that well-established fiscal policy

[12] Charles J. Hitch and Roland N. McKean, *The Economics of Defense in the Nuclear Age* (New York: Atheneum, 1969), pp. 45-46.

and bureaucratic behavior act to constrain any sharp departure from the traditional policy of balanced defense, but they do not make it impossible. Japan could certainly increase its defense budget sufficiently to put it in the class of major powers, matching mainland China or surpassing it to become Asia's number one military power by 1975 if it so chose.

Given Japan's prodigious growth rate, this would require only a gradual increase of GNP share each year of a little more than 0.2 percent, reaching 1 percent additional or a total of about 2 percent of GNP in 1975. Such a modest shift would have no appreciable effect on the growth rate and could be financed relatively painlessly by a combination of increasing tax levies, borrowing, and reallocating the budget shares within the limits to which the Japanese are accustomed. But it could hardly be done without a new sense of urgency among the general public, a revision of strategic thinking, a strengthening of the military constituencies, or a shift in political leadership marking a real departure from the past. This is particularly true since all countries with defense expenditures of this size, except for West Germany, possess nuclear weapons; and among Japanese the possession of nuclear weapons is still politically unacceptable.

An even greater shift to a policy of building to superpower status during this same period also is economically feasible, but the constraints of fiscal and bureaucratic practice would be more serious and require even greater changes in the political realm. Without such changes it must be presumed that Japan will prefer to follow its unique course of balanced defense, remaining a middle power until much further along. The crucial questions then are political.

Before turning to these, however, let us look briefly at the impact of fiscal constraints on another aspect of the defense question: arms production, purchase, or sale. Following Japan's surrender in 1945, the Japanese arms industry was dissolved, so that when the National Police Reserve was established in 1950 as a forerunner to the present Self-Defense Forces, it had to rely wholly on the United States for its arms and ammunition. Gradually over the years since then the Japanese have methodically expanded their arms production, both in quantity and quality, so that by now they are able to produce all of their own small arms, cannons, tanks, mines, submarines, and other ships. In fact, the conventional arms industry now has considerable excess capacity.

The Japanese, therefore, can be expected to develop an increasing interest in selling conventional weapons abroad. The arms industry is already pushing for this.[13] The fiscal problem is that Japan's accounts with its Asian neighbors are already heavily unbalanced. A sizable expansion of the arms trade would seem to depend on aid. Aid has not been extended for military purposes, but there is no fiscal reason not to and the aid program is destined to grow. As indicated above, the aid figure, which in 1969 stood at a level of 0.76 percent of Japan's GNP of $1.263 billion, is scheduled to be raised to 1 percent of GNP or roughly $4.7 billion by 1975. The fiscal situation would therefore permit a rather large military aid program in conventional weapons for a number of countries by 1975 if the Japanese government so chooses. As on rearmament, the constraints on arms sales are political, not fiscal.

On the other hand, Japan has not yet acquired the technological capability to produce independently from its own research and development the sophisticated fighter planes, missiles, and other electronically controlled weapons the Japanese want. For these it must still rely on licensing arrangements with the United States. It can easily afford to do this. But there are indications that Japan may be moving toward fundamental decisions to allocate much larger sums for the big science that technological "independence" would re-

[13] See Frank Langdon's discussion in Chapter IV.

quire, so that perhaps by 1975 or shortly thereafter it may be able to design its own fighters, although it may still want to import jet engines and to build its own guided missile systems. It can afford to do this too. But for technological reasons it will not be in a position to supply such military technology to others until at least the second half of the decade.

These facts also have a bearing on Japan's probable arms control policy, it being reasonable to assume that the government of Japan will be primarily interested in such measures as will enable it to keep to the methodical pace of rearmament envisaged by its traditional policy.

Vis-à-vis its Asian neighbors, this means that Japan can be expected to welcome arrangements that would seem to reduce tensions and therefore the necessity for arms and the likelihood of war; but it has no fiscal reason to fear a long-term arms race with any of them, including mainland China. It need feel no fiscal compulsion to enter agreements either for the mutual reduction of current expenditures or for freezing them arbitrarily at any particular level.

Vis-à-vis the superpowers, Japan does have a fiscal problem. It can therefore be expected during the next several decades, while Japan is building toward ultimate equality with them, that it will welcome measures to improve relations with them, particularly to relax the tensions with the Soviet Union and China. Japan can also be expected to favor arrangements that would help to close the military gap between itself and the superpowers by, for example, limiting the continuing sophistication, stockpiling, and deployment of advanced weapons, particularly nuclear—but only to the extent that such arrangements would not weaken for itself and its overseas interests the conventional and nuclear protection of the United States, limit its access to American military technology, or permanently prevent it from itself rearming to a superpower level if it wants to.

Finally, having the fiscal capacity to supply on credit to its neighbors now an increasing quantity of conventional arms and in the latter part of this decade probably sophisticated weapons systems as well, Japan's interest in restraining overseas sale of such articles for fiscal reasons can be expected to be low and to decline steadily as time passes.

CHAPTER II

Strategic Thought and the U.S.-Japan Alliance

MARTIN E. WEINSTEIN

THROUGHOUT the postwar period Japan's defense policy of minimal rearmament and American alignment has been widely misunderstood. One charge is that it represents no policy at all, but simply subservience to American demands. The other is that, however pacific Japan may appear, it cannot be trusted; as soon as it can afford to do so, Japan will surely go it alone. The truth is that Japan's defense policy is what it claims to be: strictly defensive. Moreover, it is the result of clearly-thought-out strategic conceptions, deeply held and consistently followed by its conservative leadership for more than twenty years.

The available evidence indicates that Japan's leaders formulated the basic ideas of their postwar defense policy as early as 1947 in response to the abortive effort that year by General Douglas MacArthur, the Supreme Commander of the Allied Powers for the Occupation of Japan, to conclude a peace settlement.[1] The general contended that the Occupation had accomplished its mission of "demilitarizing" and "democratizing" Japan. He argued that it was up to the Japanese themselves to give the Occupation reforms permanence, and that the greatest obstacle to that goal was the chaotic condition of their economy. He advised Washington that he could not guarantee a responsible, peace-loving, democratic Japan unless the Japanese were given a chance to regain their self-respect and provide themselves with adequate food, clothing,

[1] For a more detailed, documented account of this and other historical events mentioned in this chapter, see Martin E. Weinstein, *Japan's Postwar Defense Policy, 1947-1968* (New York: Columbia University Press, 1971).

and shelter—all items that many Japanese were doing without in 1947. MacArthur's prescription for moral and economic recovery called for the conclusion of a non-punitive peace treaty that would end the wartime blockade and readmit Japan to world trade. He favored minimal reparations and economic restrictions. He believed that with the restoration of their sovereignty and self-respect, their access to raw materials and markets, and their incentive to work, the Japanese would quickly get back on their feet economically.

Naturally enough, the leaders of the Japanese government, then headed by Prime Minister Shigeru Yoshida, were happy with MacArthur's efforts. They did, however, have serious reservations on one point. How, they wondered, was Japan's security to be protected after peace was concluded and the American Occupation forces withdrawn? The Occupation had completely abolished Japan's military forces. Under the new SCAP-sponsored constitution the Japanese people had "forever" renounced war as a sovereign right of the nation and had undertaken "never" to maintain land, sea, and air forces. Even if this constitutional ban had not existed, the terrible condition of Japan's war-shattered economy made rearmament impossible.

Japanese leaders were particularly worried because the Occupation had reduced the Japanese police also, to virtual impotence. In February 1947 a group of communist-led labor unions had attempted to bring down the Yoshida Cabinet by means of a nationwide general strike. Only the last-minute intervention of General MacArthur, who prohibited the strike and threatened to use the American Occupation forces to suppress it, saved the country from possible large-scale and insurrectionary violence. Yoshida and his colleagues in the Liberal party, as well as some of his opponents in the Democratic and Socialist parties, were convinced that unless the Japanese government was permitted to centralize and greatly strengthen its police forces before the American troops left,

Japan's parliamentary, constitutional government would be easily toppled by another general strike or a communist coup.

Yoshida and many of his colleagues also believed that this internal communist threat was related to and reinforced by an external Soviet threat. Japanese officials knew that the Soviets opposed the unified American Occupation and had wanted instead to partition Japan and have a joint occupation as in Germany. When the Occupation was taking form in late 1945, the Russians had demanded, but had not gotten, an occupation zone in the northernmost island of Hokkaido. During 1946, at the meetings of the Allied Council for Japan in Tokyo, the Soviet representative, General Kuzmo Derevyanko, made speeches bitterly denouncing American Occupation policy, insisting on an extremely harsh, punitive peace treaty, and urging the Japanese communists and socialists to unite and take radical, and if necessary violent action against their government.

The violence of Soviet propaganda attacks took on grave significance when viewed against the military dispositions. During 1946 and 1947, while the Americans were demobilizing their forces in the Far East, the Soviets kept their units intact. When the Soviets joined the war against Japan in August 1945, they captured a cluster of islands right off the northern coast of Hokkaido—Kunashiri, Shikotan, Etorofu, and the Habomai group—over which they had no previous claim, historical or otherwise. These islands are large enough for staging military operations. The Habomais are less than five miles and Kunashiri is only twenty miles from Hokkaido. Soviet air forces operating from these islands and from the Maritime Provinces could give direct support to an invasion force and were within striking distance of Japan's major industrial cities. Soviet ground units were stationed close to Japan in a wide arc extending from the southernmost Kuriles and Sakhalin, down the coast of the Maritime Provinces into north Korea.

While Yoshida was concerned over this external Soviet threat, he did not seem to have considered an invasion to be imminent. If, however, the United States were to conclude a peace treaty with Japan without the Soviet Union, and if the American Occupation forces were to be withdrawn, the Soviets were thought to be in an excellent position to assist a communist insurrection or to launch a direct attack against a weak, unarmed Japan.

Consequently, when he learned of MacArthur's peace plans in early 1947, Yoshida instructed Japanese officials to ask the American Occupation authorities how they expected Japan to be protected after the conclusion of the proposed peace treaty. The American response was vague and not very encouraging. The Japanese were told that it was still too early to worry about post-peace treaty security problems, but that when the time came, the U.S. would probably want Japan to request a security guarantee from the new United Nations.

The Japanese officials described the internal and external threats that they perceived and pointed out that the United Nations had no armed forces with which to protect Japan. As they saw it, the security question was urgent. Under instructions from Yoshida, they stated that, when a peace treaty was concluded, Japan would want to maintain its independence and protect itself against foreign invasion by concluding a mutual security agreement with the United States. Under this proposed agreement Japan would build up internal security forces capable of suppressing large-scale rioting and insurrection. The United States would guarantee Japan against invasion, would maintain its forces in the area adjacent to Japan (meaning at the time the Ryukyus and the Bonin Islands), and would in an emergency be given the use of bases in Japan itself.

In May of 1947 the Yoshida Cabinet was replaced by a coalition government headed by a socialist, Tetsu Katayama.

Despite this change, the new foreign minister, Hitoshi Ashida, followed up Yoshida's probes on the question of post-peace treaty security. During the summer and early fall he spelled out this proposal for a mutual security agreement in a number of memoranda that were sent to SCAP and to Washington.

Although more than two decades have passed since Yoshida and Ashida formulated their security policy proposals, their ideas are still highly relevant to an understanding of Japan's defense policy and to U.S.-Japan security relations, for the basic lines laid down in 1947 have continued to shape Japan's defense policy down to the present day. Yoshida and Ashida based their approach to Japan's security on two premises that are still held by the Japanese government. They calculated that the United States and the Soviet Union were the only two military powers that counted in the Far East. Despite its status as one of the Big Five, they gave little weight to China. They also assumed that the Americans and Soviets would not be able to reach a friendly understanding on the future of the Far East or Japan. Consequently, the Japanese government did not think it would be wise to attempt to provide for Japan's safety by a policy of neutrality, for Japan could be neutral only if its neutrality were respected and guaranteed by both the United States and the Soviet Union. They concluded, therefore, that Japan's security would be best served by an alliance with one of the superpowers, and their choice of the United States was almost a foregone conclusion. The Soviet Union, although situated close to Japan, could not protect Japan against the United States. The Americans were already occupying Japan. They had naval and air superiority around Japan and could protect Japan against the Soviets. Moreover, if the Japanese were ever to prosper again, they would require access to the world's sea routes leading to distant markets and sources of raw materials. Again, it appeared that only American naval and air power could guarantee Japanese access to world trade.

Two additional points in the Yoshida-Ashida security proposals need to be noted, for they have been central to Japan's postwar defense policy. First, although the Japanese government believed that both the internal communist and external Soviet threats were real, and were planning counter-measures, they were not calling for an anti-communist crusade backed by massive military forces to turn back the Red Tide. They were not even proposing that Japan be rearmed. On the contrary, the Japanese government took the position that Japan would be adequately defended against direct Soviet attack by a U.S. guarantee, by American forces stationed in areas adjacent to Japan, and by the maintenance of bases in Japan which the United States could use in an emergency. The Japanese government would handle the threat of an internal communist take-over by building national paramilitary police forces.

Secondly, neither Yoshida nor Ashida was proposing that the United States assume complete responsibility for Japan's security. They were, rather, asking the United States to guard Japan's external security while their government took care of internal security. The Japanese government, demilitarized and economically prostrate, would have their hands full protecting themselves against insurrection and infiltration. From the Japanese government's point of view, this sharing of the security burden was intended to give the proposed security agreement a mutual, reciprocal character. They believed that Japan's strategic value to the United States, its willingness to assume complete responsibility for the internal communist threat, and its desire to cooperate with the United States in defending Japan against a direct Soviet attack constituted a realistic and reasonable basis for a mutual defense arrangement. Yoshida and Ashida were planning to provide for

Japan's security by making it an ally of the United States, not a military dependency.

In 1947 the American response was still negative. In June of that year the Soviets had refused an American invitation to a Japanese peace conference. The Truman administration was preoccupied with the rapid deterioration of U.S.-Soviet relations in Europe. The prevailing view in Washington was still that the best hope for world peace and American security lay in U.S.-Soviet cooperation. At the time, the United States government opposed concluding a Japanese peace treaty without Soviet participation on the grounds that such a move would end all chances of reaching an understanding with the Russians on the shape of the postwar world. Moreover, even those in Washington who were convinced that a clash with the Soviet Union was unavoidable did not favor an early peace with Japan. They thought it might be necessary to fight the Soviets in the Far East and were in no hurry to withdraw the Occupation forces from Japan.

By 1951, however, American policy in the Far East and toward Japan had undergone a complete reversal. The Cold War was never to be more tense than it was during 1950-1951. The basic goal of American foreign policy was to hold the line against further communist expansion, either by direct or indirect aggression. The means to this goal were to strengthen the non-communist world economically and politically and to erect a barrier of regional military alliances around the Sino-Soviet bloc. In Europe, the Marshall Plan was beginning to show results and NATO was coming into being.

In the Far East, a comparable regional security system seemed absolutely imperative. In 1949 the civil war in China had ended with the communists in power in Peking and the nationalists on Taiwan. In February of 1950 the Chinese communists had concluded a military alliance with the Soviet Union, directed against Japan and the United States. In June the Soviet-sponsored North Koreans had invaded South Korea and President Truman had ordered American forces stationed in Japan to defend South Korea under UN auspices. It is worth recalling these events leading up to and into the Korean war because by January 1951, when John Foster Dulles was sent to Tokyo to negotiate in earnest with the Japanese government, the Truman administration had drawn a number of lessons from them. It had concluded that there was a Sino-Soviet communist bloc, that this bloc was intent on dominating the Far East and the Pacific, and that it would resort to military force to gain its objectives. It followed from these premises, from Japan's proximity to Korea, and from Japan's strategic value that Japan itself was in serious danger of Sino-Soviet aggression. Secondly, with NATO as a model, Dulles concluded that the best way, and perhaps the only way of containing the Sino-Soviet bloc and protecting Japan was to conclude a non-punitive peace (without Soviet participation), to permit and encourage Japan to rearm, and to enlist Japan in an American-led regional security system. In addition, Dulles was convinced that while the Occupation should be ended, extensive American forces would have to remain in Japan for the foreseeable future. Bases in Japan were vital for the support of the UN operation in Korea, and he thought that American forces were necessary to defend Japan itself.

Thus, in the four years since 1947 the Truman administration's views had shifted closer to those of the Yoshida administration, which continued to believe in a Soviet threat to Japan and wanted to counter it by entering a mutual security agreement with the United States. But differences remained. Prime Minister Yoshida still did not want Japan to rearm, and he had no intention of committing Japan to a military role in maintaining regional security. What he had wanted was a treaty providing for an American guarantee and for Japanese-American consultation and cooperation for the defense of Japan. When he met with Dulles in January 1951,

Yoshida made only one major modification of his pre-Korean war proposal. In 1947 he had favored the maintenance of stand-by bases in Japan for American use in an emergency. In 1951 he wanted the Americans to have as many operational bases as they needed to defend South Korea and Japan. For with regard to Korea, Yoshida held to an axiom that had governed Japanese foreign policy since the late nineteenth century, to wit, that in the hands of a hostile power (i.e., Russia or China), Korea is a dagger pointed at the heart of Japan. He believed that since the Japanese themselves were not able to defend their vital security interests in Korea, they ought to help the American and the UN command to do the job for them. He expected that a continued American military presence in Japan would encounter popular opposition, and he planned to sweeten the pill by placing the security agreement within the framework of the UN charter, and by providing for Japanese-American consultation on the use of American forces. In all other essential respects, however, Yoshida stood where he had four years before. Although he held to the necessity of keeping Korea from falling under Chinese or Soviet control, he did not believe that the Sino-Soviet bloc was threatening to engulf Japan and the whole Far East. He felt very strongly that Japan was not ready to rearm, either economically or spiritually. Japan could not, therefore, undertake any regional military obligations. As Yoshida saw it, the Chinese communists, lacking sea and air forces, did not greatly increase the Soviet threat to Japan. American naval and air superiority in the western Pacific was still the key to the Far Eastern balance of power and to Japan's external security. An American guarantee to use this power for Japan's defense would deter the Soviets.

Thus, the security agreement sought by Yoshida in 1951 was to be mutual and to operate within the framework of the UN charter. It was to provide for the maintenance of American forces in Japan in order to protect Japan in accordance with the charter. Japan, of course, would cooperate with the United States in its own defense. Provision was to be made for consultation on the disposition and deployment of the American forces in Japan. The treaty was to be for fifteen years.

Dulles and Yoshida were well-matched—both forceful, determined, and slow to compromise. At their Tokyo meetings in January 1951 they had little difficulty reaching agreement on the shape of a peace treaty. Yoshida was satisfied that Dulles was working toward a just, liberal settlement. Japan would lose all of its extensive prewar empire, but it would completely regain its sovereignty, unfettered by political or economic restrictions or by international supervision and control. On post-peace treaty security, however, the two men disagreed sharply. Dulles insisted that unless the Japanese government undertook to build a 350,000-man army, capable of defending Japan against a Soviet attack, and of participating in regional defense, there was no basis for a mutual security agreement. Yoshida refused to budge on the rearmament question and he would not commit Japan to a military role in regional defense. He pointed out that rearmament was forbidden by the American-sponsored constitution, and he argued that the Japanese economy could not bear the strain of rearmament. Japan's contribution to regional security should be to make itself into a stable, prosperous democracy, to cooperate in its own defense with the United States and to provide bases for support of the U.S. operation in Korea.

Not surprisingly, the negotiations quickly reached a deadlock. Ambassador Dulles could not understand why Prime Minister Yoshida did not appreciate the seriousness and urgency of the Sino-Soviet threat. Surely the war in Korea, in Japan's own backyard, should have made it clear that Japanese rearmament and participation in a regional security system were absolutely necessary to stop the Red Tide. But Prime Minister Yoshida refused to rearm.

When it appeared that the talks would collapse, one of the negotiators suggested that the deadlock be resolved by General MacArthur. Yoshida and Dulles met with the general. Dulles, to his surprise, discovered that MacArthur agreed with Yoshida. MacArthur, too, was opposed to Japanese rearmament. During the summer of 1950, when the Occupation forces were leaving Japan for Korea, he had authorized Yoshida to create a 75,000-man National Police Reserve to maintain internal security. Yoshida, it will be recalled, had been urging the creation of such a force since 1947. He was happy to see it materialize. But neither Yoshida nor MacArthur wanted to build this internal security force into a regular army. Finding himself opposed by both the prime minister and the general, Dulles had little choice but to compromise.

The final product of the Dulles-Yoshida meeting was the first U.S.-Japan security treaty, signed in San Francisco on September 8, 1951, a few hours after the signing of the peace treaty. In essence, it was a base leasing agreement, for the only substantive issue on which Dulles and Yoshida were in agreement was the necessity of having U.S. bases in Japan to support the UN operation in Korea and to protect Japan. For the rest, the treaty was an expression of intentions and expectations and read more like a joint communique than a contract. It was a provisional agreement. The United States was "willing" to keep forces in Japan, at the request of the Japanese government, in order temporarily to provide Japan with a measure of security, "in the expectation . . . that Japan will increasingly assume responsibility for its own defense against direct and indirect aggression." That was all Dulles managed to eke out on rearmament. There was no mention of a regional security system. Yoshida did not come out much better. There was only a passing reference to the UN charter in the preamble. The treaty was not mutual and made no provision for consultation on the defense of Japan or on the deployment of the American forces stationed in Japan. In fact, the treaty did even explicitly obligate the United States to use these forces to defend Japan. It only provided that: "Such forces *may* be utilized to contribute to the maintenance of international peace and security in the Far East and to the security of Japan against armed attack from without, including assistance given at the express request of the Japanese government *to put down large-scale riots and disturbances in Japan*, caused through instigation or intervention by an outside power or powers." (Author's italics.)

Mr. Yoshida fought hard to have this reference to possible American intervention to preserve internal security deleted but only managed to have it made contingent on the "express request of the Japanese government." Finally, instead of having a fixed term of fifteen years, the treaty was to be terminated only when both governments agreed that it was no longer necessary. This was a most unusual arrangement. It implied that American forces could remain in Japan even if the Japanese government decided it no longer wanted them. Of course, it also suggested that the American forces were obligated to stay until the Japanese had no further need of them, but public opinion in Japan was not likely to see it that way.

The treaty, ratified in 1952, was a temporary, minimal compromise between American Far Eastern security policy and Japanese defense policy. Japan joined the "free world" but not as a fighting member. The United States got bases in Japan that it could use to support operations anywhere in the Far East. For its part, Japan got an effective *de facto* guarantee of its own external security, in the form of the U.S. forces stationed in the country, and it also got American protection in Korea. However, the Americans failed to get their regional security system in Northeast Asia and the Japanese did not get the mutuality they wanted.

In the United States the security treaty with Japan drew little attention and was ratified without much ado, despite some concern in the Senate that we were giving the Japanese a free ride. In Japan the treaty stirred up much more opposition. The continued presence of U.S. forces in the country was bitterly criticized as a continuation of the Occupation. The provision for possible American intervention to maintain internal security and the arrangement for terminating the treaty were attacked as derogations of Japanese sovereignty. Yoshida, unable to defend the treaty as a mutual arrangement under the UN charter, made little effort to explain it to the Diet or the public. He managed to have it ratified by tying it to the peace treaty. This tactic may have appeared necessary at the time, but it led the Japanese people to believe that the Japanese government had to sign the security treaty as the price for the peace treaty. As a result, although most Japanese have continued to have a favorable opinion of the United States, they have, since 1951, viewed the U.S.-Japan security relationship as a one-sided affair, imposed on Japan against its will.

The outstanding differences between American and Japanese policy reflected in the 1951 security treaty were resolved in two stages. The first stage, from 1951 to 1957, was one of interpretation and development of the 1951 treaty. The second stage, from 1957 to 1960, was the negotiation of a new, revised treaty. Throughout, the Japanese adhered to the conceptions which from the beginning had guided Yoshida in his negotiations with Dulles. Specifically, they sought:

1. To place the security arrangements squarely within the framework of the UN charter.
2. To gain an *explicit* guarantee from the United States of Japan's external security.
3. To establish a U.S.-Japan consultative committee for the purpose of regulating the use of the American military bases in Japan and coordinating the two governments' views on questions pertaining to the defense of Japan.
4. Ultimately to replace the 1951 security treaty with a

new one which would include the above, as well as provisions for fixed duration. They were also determined in the new treaty to omit any reference to possible American intervention for maintaining internal security.

Prime Minister Yoshida and the Foreign Ministry officials concerned with security policy made a clear distinction between the first three items and the last. The former could be achieved through agreements supplementing the security treaty and would be given first priority. The latter could be realized only by making a new treaty and would have to be postponed at least until the war was settled in Korea. Although disappointed with the treaty, the Japanese officials comforted themselves that it was vague and provisional. It could be developed by interpretation and had, after all, been concluded with the understanding that eventually it would be replaced by a mutual treaty.

True, Ambassador Dulles' position had been that a new treaty would be negotiated only when Japan had rearmed and could defend itself against attack. Nevertheless, Yoshida believed that there was room for maneuver on the rearmament issue. He still considered the 350,000-man ground forces urged by Ambassador Dulles to be out of the question. He was, however, willing to expand and reorganize the National Police Reserve and to add to its mission of maintaining internal security the task of participating in Japan's external defense. His policy and that of his successors was to get the security arrangements they wanted by undertaking a minimal, limited rearmament.

Moreover, Prime Minister Yoshida believed that, given time and a closer knowledge of Far Eastern affairs, Dulles and those in Washington who thought as Dulles did would realize that the regional security arrangements appropriate to Western Europe could not be applied to Northeast Asia, that direct cooperation on security between Tokyo and Seoul was next to impossible, and that a powerful Japanese armed force and a Japanese overseas military commitment were more likely to frighten Asians than enhance their sense of safety.

Finally, Yoshida thought that Dulles had made a tactical error in the negotiations when he made large-scale rearmament the price of mutuality, for mutuality was as necessary for the United States as it was for Japan. After all, in order to make effective use of the six hundred military bases and facilities they had in Japan, the Americans were going to need the cooperation and support of the Japanese government and people. The obvious and sensible way to get this cooperation and support was to consult with Japanese officials, and persuade them that the bases were being used as the Japanese wanted them to be used.

In sum, Japan's defense policy in the 1950's continued along the basic lines set forth in 1947, as modified by the Korean war in 1950. The tasks the Japanese government set itself were to get a firm, explicit external security guarantee from the United States and to define and formalize a mutual arrangement for the defense of Japan. The means to these ends were patient diplomacy and a compromise on limited rearmament.

In April 1952, when the peace and security treaties came into effect and Japan formally regained its sovereignty, Prime Minister Yoshida announced the reorganization of the 75,000-man National Police Reserves into the National Safety Agency, to be composed of a 100,000-man National Safety Force and an 8,900-man Maritime Safety Force. These forces were not even expressly authorized to defend Japan against external attack, but during 1952 and 1953 they assumed partial responsibility for the ground defense of Hokkaido, replacing elements of the U.S. 7th Cavalry Division.

While compromising on rearmament, the Yoshida government also took steps to make the American guarantee more explicit and to develop areas of consultation with the Americans. During 1952 Soviet MIG 15 fighter planes based in Sakhalin repeatedly penetrated Japanese air-space over Hokkaido, obviously conducting reconnaissance and probably testing the U.S.-Japan security arrangements. The Yoshida government considered these overflights to be a serious threat to Japan's security. They had no means of repelling them, and they wanted the United States to take action. They did not, however, want to instigate an American-Soviet air war over Japan, nor did they want to dramatize excessively the Soviet overflights. Doing so would have undermined their argument on the limited nature of the Soviet threat to Japan and would have generated pressure for the rapid, large-scale rearmament they were intent on avoiding. As it turned out, the American reaction to the overflights was more restrained than the Japanese expected. General Mark W. Clark, commander of the U.S. Forces in the Far East, did not want to have Soviet planes inspecting American units in Hokkaido. On the other hand, because of the war in Korea he had only limited air forces available for use over Japan, and he too wanted to play down these incidents. Both governments were, therefore, happy to agree on a cautious, measured response to the Soviets. American F86 sabrejets would go up and warn the Soviets out of Japanese air-space. They would shoot only if the Soviets ignored the warnings. For the time being, the affair was to be kept out of the newspapers. The Soviet intrusions almost stopped by autumn 1952 and, following a brief flurry of overflights in December, during which a MIG was shot down, they came to a complete halt. In brief, the two governments had consulted on the defense of Japan and on the use of the American forces. The American guarantee had been invoked and had proved effective. So far, so good.

There remained, however, the possibility that the Soviets might renew their overflights, perhaps in greater force. In order to warn them that future intrusions would also be forcefully repelled and at the same time to make it clear that American military action had been requested by the Japanese government and was not in violation of Japan's sovereignty, it was decided to issue a public warning to the Soviets and to publicize the overflights. In January 1953 Foreign Minister Kazuo Okazaki and Ambassador Robert D. Murphy formally exchanged notes in which the United States agreed to Japan's request that it protect air-space under the terms of the security treaty. The notes were published in the Japanese press, along with accounts of the air action over Hokkaido during the previous year.

In a press conference a spokesman for the Japanese Foreign Ministry indicated his government's satisfaction with the way the security treaty was working. He stated that the events of 1952 and the Murphy-Okazaki notes on air defense showed, that despite the vague language of the security treaty, the Americans would defend Japan, and do so after consultation and in cooperation with the Japanese government. In fact, the spokesman said that the Murphy-Okazaki notes "set the precedent that as a matter of principle, the United States will not take military action [for the defense of Japan] unless the Japanese government requests it."

There is no record of Secretary of State Dulles' private reaction to the Japanese interpretations of the security treaty. He did not, however, publicly refute the Japanese statements. On the contrary, under Secretary Dulles' guidance, the United States government was gradually moving away from Ambassador Dulles' position of 1951.

In March 1954 the U.S.-Japan Mutual Defense Assistance Agreement was concluded. Its main purposes were to establish a legal basis for the furnishing of military equipment by the United States to Japan under the Mutual Security Act of 1951, and to clarify the terms of Japan's contribution to the support of the United States forces in Japan. But the agreement also reflected the persistence of the Yoshida government in pursuing its defense policy and growing American willingness to accept that policy. The preamble makes reference to the security treaty, stating that it was concluded in order "to promote peace and security in accordance with the purposes and principles of the charter of the United Nations." In the treaty Prime Minister Yoshida stated his commitment to a policy of limited rearmament, a policy that was to be consistent "with the political and economic stability of Japan." Lastly, as its title indicates, in the agreement the United States government acknowledged that Japan's economic and political cooperation and its limited rearmament program contributed to the common defense and constituted a proper basis for mutuality, thus conceding the point Yoshida had insisted on in the 1951 negotiations.

In the summer of 1954, his last in office, Prime Minister Yoshida again expanded Japan's armed forces and gave to them the organizational form they have retained to the present day. After a long, acrimonious debate in the Diet, his government gained approval for the Defense Agency Establishment Law and the Self-Defense Forces Law. These laws created the National Defense Agency (headed by a civilian director-general), and the Ground, Maritime, and Air Self-Defense Forces. At the time their total strength was 146,285 men. The Self-Defense Forces were authorized "to defend Japan against direct and indirect aggression and, when necessary, to maintain public order." The passage of these laws represented the Japanese government's first official acknowledgment of its responsibility for Japan's external defense. It is worth noting, however, that the Self-Defense Forces are also legally charged with defense against indirect aggression and the maintenance of public order. This was their original *raison d'être*, and it has continued to be one of their primary missions.

In December 1954 Yoshida and the Liberal party were replaced by Prime Minister Ichiro Hatoyama and the Democratic party. Within a year the two conservative parties united to form the Liberal Democratic party, which is still in power. When they took office, however, Hatoyama and his foreign minister, Mamoru Shigemitsu, were intent on creating a new foreign policy. They felt that while a friendly, cooperative relationship with the United States was politically and economically necessary, the relationship created by Yoshida was too close and confining. Japan should be more independent. The way to get greater independence was to improve relations dramatically with the Soviet Union and simultaneously to build up the armed forces. These two steps would, Hatoyama thought, reduce the Soviet threat and render Japan better able to defend itself, with less American help.

The Hatoyama approach, which promised to give greater freedom of maneuver to Japan at an early date, ought to be kept in mind, since it appears to offer what could become an increasingly attractive alternative within Japanese domestic politics. For this reason it could perhaps become more important during the next decade than it has been in the past. Like the Yoshida policy it rests on the premise that Japan's foreign and defense policies are products of U.S.-Soviet antagonism and of Japan's inability to defend itself against either of the superpowers. But unlike the Yoshida policy it prescribes the basic ingredients for a fundamental change of Japanese policy: a Japanese-Soviet rapprochement in the Far East and/or the building of Japanese forces capable of countering Soviet, if not American, military strength.

In practice, the formula proved impossible to realize in the 1950's. The Soviets refused to conclude a peace treaty with Japan or to return the northern islands unless the Japanese government abrogated the security treaty with the U.S. Prime Minister Hatoyama, eager as he was to make peace with the Soviets, would not surrender or even weaken the security treaty with the United States. Although he increased the strength of the Self-Defense Forces to approximately 180,000 men by 1956, he knew that these forces were dependent upon the United States for arms and equipment and were not adequate to offset those of the Soviet Union. He still needed the American guarantee. As a consequence, despite eighteen months of grueling negotiations, he did not conclude a peace treaty with the Soviet Union. Nor have any of his successors. He and the Soviets did, however, sign an agreement, in October 1956, formally ending the state of war between Japan and the Soviet Union and reestablishing diplomatic relations. The Soviets also agreed to support Japan's admission to the United Nations, which took place in December 1956.

Thus, despite the Japanese government's overtures to Moscow, during the two years of the Hatoyama Cabinet (December 1954 to December 1956) the U.S.-Japan security relationship continued to evolve along the lines Yoshida had hoped it would follow. Between 1954 and 1956 the U.S. forces in Japan were cut from 210,000 men to approximately 100,000. This sharp reduction in the American presence improved the tone of U.S.-Japanese relations. The curtailments were made in the aftermath of the Korean war and in keeping with the New Look strategy. Nevertheless, in Japan they redounded to Prime Minister Hatoyama's credit. And, of course, this sharp reduction in American forces, which was not accompanied by a comparable increase in Japanese military strength, meant that the Americans were, in practice, moving toward a guarantee of Japan's external defense, which is what the Japanese government wanted.

The Hatoyama government also widened the area of consultation on security matters. Specifically, they set the precedent, still in effect today, that the United States would not bring atomic or nuclear weapons into Japan without first gaining the agreement of the Japanese government. On July 28, 1955 an official spokesman in Washington announced that the U.S. Army was planning to equip its forces in Japan with Honest John missiles, capable of carrying atomic warheads. It was not clear whether the missiles would, in fact, be atomic-tipped.

The announcement produced an uproar in the Japanese press and Diet. Most Japanese, including many conservatives, vividly remembered the horrors of Hiroshima and Nagasaki. They were deeply opposed to having atomic or nuclear weapons on Japanese soil. Prime Minister Hatoyama and Foreign Minister Shigemitsu found it necessary to appear promptly before the Diet to clarify the American announcement. They explained that the deployment of the Honest Johns had been agreed upon in March 1955, in discussions between Foreign Minister Shigemitsu and Ambassador John M. Allison. The Americans had stated that they would not bring atomic weapons into Japan except in a war emergency and only with the approval of the Japanese government. Prime Minister Hatoyama solemnly assured the Diet that the Honest Johns being brought to Japan were not atomic-armed. He said, however, that while there was no present need for them, in an emergency Japan might want the U.S. to deploy atomic weapons for Japan's defense.

The Honest John incident showed more clearly than the Soviet intrusions over Hokkaido in 1953 that the use of the U.S. bases in Japan was circumscribed not only by Japanese government policy, but by public opinion as well. Apparently, the Eisenhower administration, as part of its New Look strategy, planned to offset our force reductions in Japan by equipping the remaining units with atomic missiles. When informed of these plans in the spring of 1955, the Hatoyama government demurred, arguing that the introduction of atomic weapons into Japan would turn public opinion against the United States and the security arrangements. The Americans seem to have replied that the NATO forces were already equipped with atomic weapons, and that they were a strategic necessity in Japan. The upshot, again, was a compromise. The Hatoyama government agreed to have the Honest John missiles and crews set up in Japan without atomic warheads. If the international situation should deteriorate to the point where both governments believed atomic warheads were needed, then the Japanese government would agree to their being brought into the country.

The vagueness of the American announcement as to whether the missiles were atomic-armed was dictated partially by security regulations and partially by the desire to test Japanese public opinion. As it turned out, the anti-atomic furor was even greater than the Hatoyama government had expected. As a result, the U.S. government came to recognize that its bases in Japan would have to be limited to a conventional, supporting role in its Far East strategy. The necessity for close consultation and cooperation on the use of the bases was reaffirmed. Not only would the Japanese government be consulted on the deployment of the American forces stationed in Japan, but it would also have a voice in deciding the weapons with which they were to be equipped.

The Hatoyama government, in its effort to improve Japan's international status, also tried to replace the 1951 security treaty with a more mutually explicit agreement. On August 29-31, 1955 Foreign Minister Shigemitsu visited Washington to clarify Japan's Soviet policy and to set forth Hatoyama's desire for a new treaty. The United States, as we have seen, had been quite liberal in interpreting the treaty, but formal revision proved to be another matter. Secretary Dulles again insisted that if the Japanese wanted a mutual treaty, they would have to rearm to the point where they could defend themselves against a Soviet invasion, and they would have to be willing to share with the United States the burden of maintaining regional security. In view of the armistice in Korea and changes in weaponry, Secretary Dulles thought that Japanese ground forces of about 200,000 men, rather than 350,000 men, would be adequate to protect Japan. At the time, the Ground Self-Defense Force numbered approximately 150,000 men. Shigemitsu indicated that his government planned to increase the ground forces to 180,000 men by 1958 and urged that negotiations for a new treaty be started promptly, on the assumption that this figure would be realized.

Dulles then raised the question of Japanese participation in regional defense. In an apparent softening of his earlier arguments, he told Shigemitsu that he was not, for the moment, asking Japan to commit itself to the defense of the Far East (South Korea and Taiwan), but he thought it only reasonable that they share in the defense of the western Pacific, including, for example, Guam. In his eagerness to get negotiations started for a new treaty, Shigemitsu decided to risk a political storm at home.

In return for an American promise to negotiate the 1951 security treaty with one of greater mutuality, he agreed in a joint communique that Japan would rapidly strengthen its defense power to the point where it could: ". . . assume primary responsibility for the defense of its homeland and be able to contribute to the preservation of international peace and security in the Western Pacific."

When the communique was made public in Tokyo in early September 1955, it caused a more serious political storm than the Honest John missiles had in July. Prime Minister Hatoyama and Foreign Minister Shigemitsu decided that the "Western Pacific" commitment was too hot to handle. They promptly dropped it, explaining to the Diet and the press that in the communique Japan had assumed no overseas commitments, military or otherwise. The paragraph at issue, they said, was simply an agreement in principle. Following this fuss, the Hatoyama Cabinet shelved its plans for treaty revisions.

In retrospect, it appears that the efforts of the Hatoyama Cabinet to alter Japan's foreign and defense policies and to improve dramatically Japan's international position were only moderately productive. Despite the normalization agreement with the Soviets, political and economic relations remained extremely limited. Entrance to the UN was a necessary step in Japan's recovery, but it did not appreciably affect Japan's international status. Although Hatoyama began his prime ministership resolved to give his country a more independent position by reducing the importance of the U.S.-Japan security relationship, he soon realized that the alternatives to this relationship—trusting the Soviets or massive rearmament—were neither practicable nor desirable. By the end of his term of office, it was clear that what had begun as a repudiation of the Yoshida policy had developed, instead, into a variation of it.

Tanzan Ishibashi, who followed Hatoyama, was in office only two months when illness forced him to resign. He was succeeded by Nobusuke Kishi (February 1957 to July 1960). Kishi's foreign policy was a conscious return to the Yoshida approach. He did not extend the Hatoyama effort for a rapprochement with the Soviets. He was content to seek security, prosperity, and international status for Japan in partnership with the United States as a loyal member of what was then called the Free World.

In defense policy also, Kishi hewed close to the Yoshida-Ashida line. He had accompanied Foreign Minister Shigemitsu to Washington in 1955 and had not forgotten the fate of the "Western Pacific" communique. Under his government there would be no further hints that Japan might assume a military role in maintaining regional security. He would continue gradually to build up the Self-Defense Forces, and would succeed in negotiating a new security treaty. But for him as for Prime Minister Yoshida, mutual security meant cooperating in the defense of Japan.

In June 1957 Prime Minister Kishi visited Washington. The main purpose of his trip was to get negotiations started for a new security treaty. He took with him plans for moderate expansion and improvement of the Self-Defense Forces but otherwise offered no concessions. Rather, he emphasized Japan's growing prosperity, stability, and status, the lessening of Cold War tensions, and the provisional nature of the 1951 security treaty. He took the position that the 1951 treaty, based as it was on the notion that Japan had no means to defend itself and that the United States would defend Japan as it saw fit, was clearly out of date. The treaty no longer described the actual security arrangements and was a continual source of controversy in Japan. A new, mutual treaty would be in the interests of both governments and should be promptly negotiated. Prime Minister Kishi urged that the new treaty should: (1) be placed clearly within the framework of the UN charter; (2) provide for regular, formal consultation on the equipment and deployment of U.S. forces stationed in Japan; and (3) be made effective for ten or fifteen years, after which the treaty could be extended or terminated by either party.

By this time President Eisenhower and Secretary Dulles were ready to agree with most of what Prime Minister Kishi had to say. Japan, it was true, had not rearmed to the level they desired, nor was it offering to assume a military role in regional security. But neither had the NATO members rearmed adequately and, following the 1956 Suez crisis and the rift between the United States, Great Britain, and France, regional security may have lost some of its magic. Moreover, Japan had staged an extraordinary economic recovery and had become a more stable, self-reliant, and cooperative ally than had been expected in Washington. Dulles could assure himself that the 1951 treaty had been a good piece of work at the time. But he himself had insisted that it should be provisional, and now it needed revising. Finally, Dulles was still hoping that the Japanese, given a new, mutual treaty, might of their own volition take a more active, responsible role in maintaining the security of the Far East.

In the joint communique issued at the close of this 1957 meeting, Eisenhower and Kishi announced the opening of a "new era" in Japan-American relations, based on equal partnership and close cooperation. They gave substance to these words by establishing the United States-Japan Committee on Security, composed of the Japanese foreign minister and director-general of the Defense Agency, and the U.S. ambassador to Japan and the commander-in-chief, Pacific. Questioned by the press, Dulles explained that the committee was authorized to discuss all questions relating to the implementation of the 1951 treaty, including its revision.

The 1957 Eisenhower-Kishi meeting marked the end of the development of the security treaty by interpretation and implementation. The Japanese government had created a Self-Defense Force of 210,603 men, including ground forces of about 170,000 men. These figures suggest that the Japanese had come much closer to developing the kind of forces the United States had insisted were necessary than was actually the case. Both governments knew that the Self-Defense Forces were not adequately equipped or properly deployed to prevent a Soviet attack. Only four of the ten understrength divisions of the Ground Self-Defense Force were in Hokkaido, and two of these divisions were training units. Most of the remaining six divisions were stationed not on the Japan sea coast facing Asia, but close to the major population centers of Honshu and Kyushu, where they could best maintain internal security, if it became necessary to do so. For, despite Japan's economic recovery and apparently stable representative parliamentary political system, the conservative government still believed that there was a real threat of a communist-led coup or a general strike that might lead to insurrectionary violence. The Maritime Self-Defense Force was a coast guard rather than a naval force. It could prevent infiltration but could do little to oppose an armed attack or to protect Japanese shipping. The Air Self-Defense Force, although it was to grow much stronger during the coming decade, was in 1957 still a reconnaissance rather than a fighting force. Thus, the basic operational mission of the Self-Defense Forces was still the maintenance of internal security. They were not prepared to assume the primary responsibility for external defense.

On the other hand, almost all American combat infantry units had been withdrawn from Japan. The remaining 77,000 men were manning naval and air bases and army supply depots. Their mission included cooperation in the defense of Japan but was primarily to provide logistical support for American military forces in South Korea and elsewhere in the Pacific and the Far East. In practice, therefore, Japan was being protected from external attack not by Japanese or American forces in Japan itself, but by the American divisions in South Korea and by our naval and air forces in the western Pacific. The 1951 security treaty had become a symbol of the American commitment to help defend Japan. In short, it was developing into the guarantee that Yoshida and Ashida had wanted.

The issue of mutuality was working out as the Japanese had hoped, for Japan, still not able to defend itself, was obviously not moving toward a military role in defending the Far East. Although Dulles, in his enthusiasm for regional security pacts, had succeeded in creating SEATO for Southeast Asia, he had learned that the disputes and complexities which characterized the relations between Japan, South Korea, and Taiwan made it impossible for them to form a Northeast Asia treaty organization. Nevertheless, the Murphy-Okazaki notes on air defense, the Mutual Defense Assistance Agreement (MDAA), the Shigemitsu-Allison understanding, and, finally, the formation of the U.S.-Japan Committee on Security all showed that the U.S. government was, in practice, willing to consult and cooperate with the Japanese government in the defense of Japan. And we should also note that the MDAA and Japan's joining the UN had moved the security arrangements into the framework of the UN charter.

Apparently, Yoshida and his colleagues had calculated correctly in 1951. Limited, gradual rearmament and a patient, skillful diplomacy had led the Americans toward acceptance of Japan's defense policy. This is not to suggest that the Japanese policy-makers had been gifted with perfect foresight or that the U.S. government had become converted to Japanese strategic perceptions and defense policy. What had happened was that the stalemate in Korea, the development of nuclear missiles, changes in Soviet policy, and the New Look strategy —all factors beyond the control of the Japanese government—were gradually bringing American Far Eastern security policy closer to Japanese defense policy.

The negotiations for the 1960 Treaty of Mutual Cooperation and Security were conducted in fits and starts through 1958 and 1959, principally in Tokyo, by Prime Minister Kishi, Foreign Minister Aiichiro Fujiyama, and Ambassador Douglas MacArthur III, the general's nephew. The delays in the negotiations stemmed principally from domestic political problems in Japan rather than from differences between the two governments over the substance of the new treaty. For as we have seen, the purpose of the new treaty was not to alter the U.S.-Japan security relationship but to set down in writing the actual arrangements that had evolved since 1951.

The root of Kishi's domestic difficulties was that he had a narrow, unstable base of support in his own Liberal Democratic party. He had, therefore, to devote an unusual amount of time and energy to building and holding together the uncertain factional alliances upon which his premiership and his power to conduct foreign policy depended. This is, of course, a problem common to all governments. It is, however, worth taking note of Kishi's difficulties because two of the more important issues that bedeviled him were directly related to defense policy.

The first of these issues had to do with the authority of the Japanese national police to maintain internal security. In the fall of 1958, just when the treaty negotiations were getting underway in Tokyo, Kishi attempted to have the Police Duties Performance Law amended by the Diet in order to prevent and control labor disorders and violence. Much to his dismay, the vague language of the proposed amendment, which seemed to give the police unnecessarily wide powers reminiscent of those they exercised in prewar Japan, provoked widespread criticism. A nationwide anti-police law revision movement sprang up with unexpected speed and force. On November 5-7, 1958, four million Japanese demonstrators staged strikes across the country protesting against the amendment.

Sensing an opportunity to depose Kishi and strengthen their own political positions, a number of powerful faction leaders in the Liberal-Democratic party who had previously supported the amendment publicly denounced the prime minister for his "high-handed approach" to police law revision. Faced with this dissension in his party, Kishi backed away from the amendment, suspended substantive negotiation on the treaty, and devoted himself through the winter and spring of 1959 to mending his political fences in preparation for a House of Councillors election scheduled for June 1959. Kishi's efforts at police law revision show that he, as well as many other conservatives, were in 1958 still deeply concerned over the threat to internal security posed by the Japanese communists and radical socialists, whose leaders denied the legitimacy of the elected government and were preaching revolution. Moreover, the demonstrations and strikes against the proposed amendment highlighted the willingness of a surprisingly large number of Japanese to express their opposition to the government by taking to the streets, giving Kishi a foretaste of the much larger and more violent opposition he was to encounter in 1960 when the new treaty came up for ratification.

The second issue that embroiled Kishi within his own party concerned the area to which the new treaty would apply. As chance would have it, this question first arose in the negotiations in the fall of 1958 just when police law revision was diverting Kishi's attention from foreign policy. As we have noted, the Eisenhower administration was willing to conclude a mutual agreement providing for consultation and cooperation for the defense of Japan. Ambassador MacArthur indicated, however, that his government still believed that the security of Japan was inseparable from the security of the Far East and expected that Japan would cooperate in some manner in regional defense. Kishi was willing to *consult* on regional security but not to commit Japan to a *military role.*

During the discussion of the treaty area, the status of the Ryukyu and Bonin Islands entered the negotiations. Kishi, it seems, was toying with the idea of taking some responsibility for their defense. Until the United States captured and garrisoned these islands in 1945, they had been part of metropolitan Japan. In Article 3 of the San Francisco peace treaty, Japan gave to the United States "all and any powers of administration, legislation and jurisdiction over the territories and inhabitants of these islands." During the 1950's it was agreed that under Article 3 Japan retained residual sovereignty over the islands. This meant that the United States could use the islands as it saw fit for the duration of the Cold War, but that when a relaxation of international tension permitted, the islands would be returned to Japan. The Japanese government agreed to this formula, knowing that the American bases and nuclear weapons on Okinawa were playing a vital role in Japan's defense and preferring to give the United States a military carte blanche on Okinawa while insisting on limiting its use of bases in Japan proper. Kishi was willing to continue this arrangement, but he wanted to reassert and strengthen Japan's claim to residual sovereignty and thought that he might do so by agreeing to cooperate with the United States in the defense of the islands. The conservative faction leaders who were attacking Kishi over the police law thought they saw another issue on which to discredit the prime minister. They announced with great fervor that Japan should accept no responsibility whatsoever for the defense of the Ryukyu and Bonin Islands until they reverted to Japan, hinting that they favored prompter reversion than did Kishi. During the winter and spring of 1959 Kishi quietly adopted this argument as his own.

The hullabaloo in the LDP and in the Japanese press over Okinawa contrasted sharply with the quiet treatment of the issue in the governmental negotiations. The U.S. government was perfectly willing to reaffirm Japan's residual sovereignty. Neither government was especially eager at that time to share the defense of Okinawa. This was not the last time that Okinawa would stir up a noisy and emotional political controversy in Japan while remaining a relatively subordinate if not unimportant issue in the relations between the U.S. and Japanese governments.

The upshot of this domestic political situation was that despite the rumblings of popular discontent and the bitter intraparty squabbling, the Liberal Democrats did well in the election of June 1959. Prime Minister Kishi interpreted the election results as a vote of confidence in his leadership. He reshuffled his cabinet during the summer and in the fall of 1959 again turned his attention to the negotiations. A new treaty was quickly hammered out.

The U.S.-Japan Treaty of Mutual Cooperation and Security, signed in Washington, D.C., on January 19, 1960, represents the realization of all the items sought by Yoshida in 1951. In the preamble and in Articles 1, 5, and 7 the security arrangements are placed squarely within the framework of the United Nations charter. Articles 2 and 4 make provision for cooperation and mutual aid. In Article 4 it is agreed that: "The parties will consult together from time to time regarding the implementation of this Treaty, and at the request of either Party, whenever the security of Japan or international peace and security in the Far East is threatened." In Article 5: "Each Party recognizes that an armed attack against either Party in the territories under the administration of Japan would be dangerous to its own peace and safety and declares that it would act to meet the common danger in accordance with its constitutional provisions and processes."

It should be noted that while the Japanese government is obligated to consult on security matters pertaining to the Far East, it is obligated to act only in the event of an armed attack against Japanese territory. These provisions fulfill the Japanese concept of mutuality, while inclining only slightly toward the American notion of regional defense. Moreover, it should be noted that under the 1960 treaty there is no authorization, as in the 1951 treaty, for American intervention to maintain Japan's internal security.

Article 5 also disposed neatly of the question of the defense of the Bonin and Ryukyu Islands. Japan would not become responsible for their defense until they were returned to Japanese administrative control. On the other hand, the United States would then continue to be responsible for their defense as part of Japan.

There is no mention of Okinawa in the treaty, but in an Agreed Minute to the treaty Prime Minister Kishi expressed his government's concern over the welfare and safety of the islanders, on the grounds that "Japan possesses residual sovereignty over these islands."

In Article 6 it was agreed that: "For the purpose of contributing to the security of Japan and the maintenance of international peace and security in the Far East, the United States of America is granted the use by its land, air and naval forces of facilities and areas in Japan." It should be noted that under the 1960 treaty the United States is granted the use of bases but is not required, as it was in the 1951 treaty, to keep forces in Japan.

In an exchange of notes it was made clear that in the implementation of Article 6: "Major changes in the deployment into Japan of United States armed forces, major changes in their equipment, and the use of facilities and areas in Japan as bases for military combat operations to be undertaken from Japan other than those conducted under Article 5 of the said Treaty, shall be the subjects of prior consultation with the Government of Japan." The prior consultation notes provide that American forces in Japan will not be reinforced or withdrawn, that nuclear weapons will not be brought into Japan, and that the U.S. forces in Japan will not be sent to fight outside Japan, unless the Japanese government first approves of these actions.

On the questions of duration and termination, Article 10 provides that: ". . . after ten years, either Party may give notice to the other Party of its intention to terminate the Treaty, in which case the Treaty shall terminate one year after such notice has been given."

If we view the 1960 treaty in terms of the ideas formulated by Yoshida and Ashida in 1947, modified by Yoshida in 1951, and then pursued by him, Hatoyama, and Kishi through the 1950's, we can only conclude that it represents the realization of Japan's defense policy and was a major diplomatic achievement for the Japanese government. And yet, the 1960 treaty proved to be Prime Minister Kishi's undoing. Instead of receiving popular acclaim for his diplomatic success, he was driven from office in the summer of 1960 by the large-scale protests, strikes, and rioting against him and the new treaty that occurred in May and June. These difficulties led President Eisenhower to cancel a scheduled visit to Japan, which was intended to coincide with the ratification of the treaty. In brief, although the 1960 treaty was a diplomatic success, it was a disaster in Japanese domestic politics and at the time threatened to weaken rather than to improve Japanese-American relations.

This fascinating paradox can be understood in terms of two basic facts of Japanese political life. The first is that until very recently every postwar Japanese government has given top priority to economic policy and domestic political stability and has purposely played down foreign and defense affairs. For the most part this approach has been in keeping with the attitude of Japanese voters. As a result of the terrible suffering and traumatic defeat in the Pacific war, they simply have not wanted to bother with international affairs. To the extent that they have given any thought at all to foreign and defense policy, they have inclined toward a vague pacificism and isolationism. This goes a long way toward explaining why the conservatives have been able to remain in power continuously throughout these years in spite of the fact that most Japanese voters have been dissatisfied in varying degrees with their government's foreign and defense policies, and despite the opposition parties' strenuous efforts to bring down the government on foreign policy issues. But the relative indifference of the voters and the government's conscious catering to and use of this indifference have led to a dismal lack of popular understanding of foreign and defense issues. In short, the popular opposition to the 1960 treaty was largely a product of the Japanese public's ignorance and misunderstanding of the treaty and of defense policy. And the government, beginning with its handling of the 1951 treaty, had contributed to this ignorance and misunderstanding.

A second factor that helps to explain the disparity between the Japanese government's diplomatic success and its domestic difficulties over foreign policy is the treatment given to foreign and defense affairs by the Japanese mass media. Throughout these last two decades the press has tended to be deeply suspicious and highly critical of the government's foreign and defense policies. This attitude in large part has stemmed from a sincere anxiety on the part of journalists that the conservatives, unless carefully watched, would use the U.S.-Japan relationship as a pretext for once again making Japan a militarist, authoritarian state. As a result, however, the mass media in Japan have shown a persistent and often indiscriminate and unfair bias against government policy. And the leaders of the Japanese government, unlike their American counterparts, have never challenged the right of the press to criticize them unfairly.

From the point of view of American Far Eastern policy, the 1960 treaty was a satisfactory arrangement. By 1960 the U.S. government was no longer pressing for large-scale Japanese rearmament or a Japanese military role in maintaining regional security. There was, therefore, nothing we could hope to gain by refusing to acknowledge openly our guarantee of Japan's external defense and the mutuality of our relations. On the contrary, the United States government negotiated the new treaty with the intent of confirming, strengthening, and stabilizing the security relationship that the Japanese government had been seeking all along. In brief, by 1960 the Eisenhower administration was aware of the Sino-Soviet rift. The communist military threat in the Far East had clearly subsided since the anxious days of the Korean war. It appeared that the American combat forces in South Korea, on Okinawa, and in the western Pacific were adequate to stabilize the Far East. It was, after all, enough to have a prosperous, stable Japan as an economic and political ally and to have bases and logistical support available in Japan in the event that they were again needed.

It seems fair to conclude that American policy toward Japan in the 1950's, culminating in the 1960 treaty, was flexible, realistic, and prudent. Dulles' policy toward Japan suggests that he was not as doctrinaire or as rigid as he himself sometimes wanted to appear. He seems always to have appreciated the necessity of a cooperative, firm relationship with Japan. The uncertainty of the American military position in Korea in 1950-1951 led him to demand more of the relationship than Yoshida was willing to give. Following the Korean armistice, however, Dulles was willing to retreat from his stand in the 1951 security treaty negotiations and to move toward Yoshida's policy. He seems to have perceived that the security arrangements that Mr. Yoshida wanted were well designed to serve American as well as Japanese interests.

Following the cancellation of President Eisenhower's trip to Japan and the demonstrations and riots against Prime Minister Kishi and the 1960 security treaty, one might expect that the Liberal Democrats would have been voted out of power and that the Socialists, who had consistently opposed the security relationship with the United States, would have formed a government and set Japan on a course of unarmed neutrality. In fact, nothing of the sort occurred. Instead, in July of 1960 his fellow Liberal Democrats pressured Kishi into resigning and selected Hayato Ikeda to replace him as prime minister.

Prime Minister Ikeda took the position that the new security treaty was satisfactory and that the unrest and violence of the spring had been directed against Kishi's chilly, arrogant manner rather than against the party, its program, or its foreign policy. He also believed that much of the popular frustration and irritation vented against Kishi had resulted from the prevalent feeling that Japan was a helpless pawn in the Cold War. Ikeda projected a warmer, more soothing image than had Kishi. Upon taking office, his immediate tactic was to get the voter's mind off foreign policy and the treaty by drumming up popular interest and enthusiasm behind his ten-year income doubling plan. By December 1960 he felt ready to face a general election, the results of which indicated that his diagnosis and prescription had been correct. The Liberal Democrats won handily in a peace-and-prosperity campaign.

The foreign and defense policies of the Ikeda government (July 1960 to November 1964) were essentially the same as those of its predecessors. Japan's foreign policy continued to be built around its relations with the United States and its reliance on the U.S. defense guarantee, as expressed in the new treaty. Despite the widening Sino-Soviet rift and China's increased bellicosity, the Ikeda cabinet believed that the Soviet Union continued to be the main threat to Japan's security. The program for the gradual build-up of the Ground Self-Defense Forces was modified to increase their effectiveness in maintaining internal security. Their role in defending against external attack was, together with the Air and Maritime Self-Defense Forces, to hold off an attacker until powerful American units could enter the battle. In effect, under Ikeda the function of the Self-Defense Forces became the strengthening of the American guarantee. The Self-Defense Forces perform this mission by being able to defend against low-level, probing attacks, thus assuring that if and when a Soviet initiative is taken, it will have to be on a large enough scale to call for a strong, quick American response. That is still their function today.

Although he did not change Japanese foreign and defense policy, Hayato Ikeda did strive to give the Japanese people a greater sense of status and prestige in international affairs. He told the voters again and again that Japan, along with Western Europe and the United States, was one of the "three pillars of the Free World." He skillfully used the 1964 Tokyo Olympic Games to create in his constituents the confidence that they were a prosperous, progressive, and peaceful nation. On the whole, however, Ikeda stuck to generalities in discussing foreign policy. He avoided specific, potentially controversial issues and put most of his time and energy into domestic economic affairs.

The Kennedy administration's reaction to the protests and riots accompanying the Japanese ratification of the new treaty was to attempt to dispel the misunderstandings in Japan that had produced the outbursts of the spring of 1960. As his ambassador to Japan, President Kennedy wisely chose Professor Edwin O. Reischauer, a man with extensive experience in and knowledge of Japanese affairs, who was well known in both intellectual and political circles in Japan. Ambassador Reischauer's main task was to resume the broken dialogue with Japan by clearing up the distortions and confusions that had grown in the public mind around the U.S.-Japan relationship. It was not an easy task. Of course, the Ikeda government wanted friendly, close relations with the United States, but they were sometimes concerned that the ambassador would clarify matters they wanted left unclear. One of the basic elements of the conservative domestic political strategy, it will be recalled, has been to focus public attention on domestic, pocketbook issues while playing down foreign policy, including relations with the United States. The opposition and the mass media have always been happy to talk about U.S.-Japan relations, but usually in a highly critical vein, and they occasionally tried to use Ambassador Reischauer as a weapon against the Ikeda Cabinet. Reischauer, aware of these obstacles and pitfalls, went energetically yet prudently about his job. He maintained close, friendly official relations, let the opposition and the media know what American policy was, and made frequent public appearances that improved the Japanese image of the United States. These American efforts, together with the success of Ikeda's peace-and-prosperity politics, had the desired effect. By November 1964, when Ikeda left office, U.S.-Japanese relations were closer, more mutual, and more equal than at any time since the end of World War II.

Although the 1960 treaty represented the realization of the basic goals of the Japanese government's defense policy and provided a realistic, sound basis for the continuation of the U.S.-Japan security relationship, it did not solve all the questions relating to security policy between the two governments.

During Prime Minister Eisaku Sato's premiership, which began in November 1964, three significant issues have come to the fore. These have been: (1) the size and visibility of the U.S. forces and bases in Japan; (2) Okinawa; and (3) the status of the 1960 treaty following the expiration of its fixed, ten-year term in June 1970.

The first point to be made about the American forces and bases in Japan is that the issue between the two governments has *not* been whether the Americans should stay or go. The Japanese government has wanted American forces in the country. However, in keeping with its conception of the defense guarantee and in order to avoid giving the opposition an issue, they have wanted to reduce the number of American bases and to move the Americans out of the congested, politically hostile cities into the less crowded countryside. Throughout the 1960's the American government's military strategy and its desire to maintain good relations with Tokyo have led it to move along the lines urged by the Japanese government.

As we have noted, there were over 200,000 American troops and approximately 600 military installations in Japan at the end of the Korean war. By 1960 all American combat infantry units had been removed from Japan under the New Look strategy. The remaining forces, principally air and naval units, totalled approximately 50,000 men stationed at about 250 bases. Most of these 250 "bases" were supply and communication facilities, often a small cluster of warehouses or a single radar tower.

During the 1960's the Kennedy and Johnson administrations dropped the New Look, with its reliance on strategic, retaliatory forces. They built up American conventional, tactical forces, including ground forces, in an effort to give us a greater capability to fight in limited, local wars. The purpose of these changes was to avoid putting the U.S. government in a position where it would have to choose between using nuclear weapons to fulfill a defense commitment and not fighting at all, thereby defaulting on its commitment. This "flexible, graduated response" strategy, however, did not imply large-scale or permanent deployment of American ground forces to overseas bases. It meant, instead, building units that could fight anywhere in the world, and providing air and sea transport to move them rapidly and support them in combat. Consequently, during the past decade American troop strength fluctuated in Europe depending on how the Soviets behaved, rose sharply in Southeast Asia in order to fight in Vietnam, and dwindled in Japan and in South Korea. At the end of 1969 there were only about 40,000 American servicemen in Japan, and the number of bases was reduced to less than 100. Moreover, with few exceptions, the remaining American forces are out in the countryside, where they create fewer safety hazards and political complications.

It seems fair to conclude that the base issue has been dealt with satisfactorily, but that it still has potential for creating future difficulties. The opposition parties in Japan have not been able to make much political capital of it, but they hang onto it. The Japanese government has been content that the American units in Japan are strong enough to guarantee Japan's defense and to support the UN command in South Korea but not so powerful as to create unnecessary tensions in the Japan area. The U.S. government, for its part, has been satisfied that the defense arrangements for Japan and South Korea have contributed to the stability and peace of Northeast Asia, and that the logistical and support bases in Japan have enabled American naval and air units to back up its defense commitments throughout the Far East.

The Okinawa issue has been related to the bases issue, and like the base issue it has generated much more noise and excitement in Japanese domestic politics than in the relations between the American and Japanese governments. This is not unnatural or reprehensible. The Japanese government, since the middle 1950's, has been willing to sacrifice the aspirations of the one million Japanese people on Okinawa in the interest of its defense policy. As we have seen in our discussion of the 1960 treaty, the Japanese government has wanted the United States to keep nuclear weapons on Okinawa and to use its Okinawan bases for the defense of South Korea, Taiwan, and Southeast Asia because they believed that these activities were necessary for Japan's security and that it was better to have them conducted on Okinawa than in Japan proper. The desire of the Okinawans to escape their quasi-colonial status and to once again be part of Japan was disregarded. This state of affairs was bound to stir up a controversy in Japan, and beginning in 1967 it did.

Early in the year the Japanese ambassador in Washington made a speech, which received great attention in Japan, in which he stated that if the Japanese people wanted Okinawa they would have to take it back with the American nuclear weapons still on the island. The opposition parties were quick to charge that Prime Minister Sato wanted to introduce nuclear weapons into Japan itself. They demanded the immediate reversion of Okinawa, the removal of all U.S. bases from Okinawa and Japan, and the abrogation of the security treaty. The mass media fastened onto Okinawa, and by spring of 1967 it had become a major public issue.

It is not clear to what extent Prime Minister Sato himself created the Okinawa issue, but once it got in motion he had no choice but to hang on to the tiger's tail. He promptly backed away from his ambassador's position on the nuclear weapons. At the same time, he assured the voters that he would get Okinawa back as soon as he could persuade the United States to give it up, and in an effort to make room and gain time for maneuver on Okinawa he also proposed that the strategically less important Bonin Islands be promptly returned to Japan. He avoided taking a stand on the Okinawa bases and nuclear weapons, knowing that his own Liberal-Democratic party was deeply divided on these questions. Some of his associates insisted that the American nuclear missiles on Okinawa were indispensable to Japan's defense and that Japan itself ought to think of becoming a nuclear power. Other conservative leaders felt that Okinawa could be safely denuclearized and that Japan's security required only that the United States retain conventional bases on the island, under the same prior consultation restrictions as those on the Japanese mainland. The latter position became known as the "mainland formula."

There were also differences of opinion over Okinawa within the Johnson administration. It was generally agreed, however, that before our government could decide on an Okinawan policy, we would have to know just what the Japanese proposed to do about the bases and the nuclear weapons. This was the state of affairs when Prime Minister Sato met with President Johnson in Washington in November 1967, ostensibly to settle the Okinawan question. The prime minister was not quite sure what to do about Okinawa, except that he had to show some progress toward reversion. Thus, in their joint communique the president and the prime minister agreed that the Bonins would be returned during 1968 and that a solution for Okinawa would be worked out in a few years. In Japanese, "a few years" was translated into two or three years, and back in Tokyo Sato used this vague promise to hold off the opposition and the press with one hand while with the other he worked hard to pull his own party together.

By the summer of 1969 he had built a consensus. In November of that year, when he met with President Nixon, it was agreed that Okinawa would be returned to Japanese administrative control during 1972, under the "mainland formula." It should be noted in passing that, contrary to American hopes, Sato did not promise Japan's adherence to the Nuclear Non-Proliferation Treaty. This was probably because his own hard-liners, while willing to give in on Okinawa, were not ready to give up Japan's option to become a nuclear power.

The question of what to do about the 1960 security treaty after June 1970, even more than the base issue or Okinawa, has been almost entirely a matter of Japanese domestic politics. The U.S. government has not seriously considered terminating or modifying the treaty. It has consistently taken the position that the treaty contributes to the peace and stability of Asia and that it should be extended, as it stands, indefinitely into the future. Our government has contended that the provisions of the treaty are sufficiently vague and general to accommodate almost any specific defense arrangements that we and the Japanese government may want to make in the future. For example, as we noted in discussing Article 6, it is entirely possible to continue the treaty without having a single American serviceman stationed in Japan. Finally, our government pointed out that formal revision of the treaty would mean that it would have to be put through the U.S. Senate and the Japanese Diet.

Since the opposition parties in Japan would like nothing better than to get another chance to attack the treaty in the Diet, one might suppose that the Japanese government and the Liberal Democrats would have been intent on avoiding revision. In fact, however, between 1965 and 1968 there was a factional dispute in the party and the government over whether the treaty should be automatically extended for an indefinite period, or revised and extended for a fixed term. The proponents of the latter view argued that a revised treaty extended for another ten years would be a firmer arrangement and would not be subject to sudden abrogation if, for instance, an anti-treaty government were to assume office in either country. By June 1968, however, automatic extension had become the policy of the LDP and the government.

Consequently, Prime Minister Sato and President Nixon,

in their joint communique of November 1969, "affirmed the intention of the two Governments firmly to maintain the treaty." This means that although both governments remain free under Article 10 to abrogate the treaty by giving one year's notice, both governments expect the treaty to continue well into the 1970's.

When Prime Minister Sato returned to Japan from Washington with the joint communique, he promptly called a general election for late December 1969. In the campaign, he asked for a popular mandate for his policy on Okinawa and the security treaty. The vote was a stunning victory for Sato and the LDP.

This review demonstrates that since Japan regained independence in 1952, its security policy has been guided not only by a well-thought-out fiscal policy of balanced defense, but also by a strongly and consistently held core of strategic thought and military policy. With the partial exception of the Hatoyama Cabinet (1954-1956), all of Japan's prime ministers from Shigeru Yoshida down to Eisaku Sato have made defense policy in the belief that Japan cannot defend itself militarily against either the Soviet Union or the United States. They have perceived the United States and the Soviet Union as competing superpowers in the Far East. They have understood that whichever of the superpowers has access to the Japanese islands and to Japanese skills and resources would enjoy a decisive military advantage in the western Pacific and on the periphery of Northeast Asia. Given these beliefs and perceptions, they have concluded that Japan can best provide for its external military defense by allying itself with one of the superpowers. The conservative leaders have rejected the Socialist policy of unarmed neutrality on the grounds that in the pinch the superpowers could not be trusted to respect that neutrality and that, unarmed, Japan would not have the means to protect its neutrality.

The basic questions with which they have dealt, therefore,

have been: Which of the superpowers to choose as an ally? And what kind of alliance relationship to build?

The answer to the first question was easy for Prime Minister Yoshida to make, and easy for his successors to accept. The United States, and not the Soviet Union, occupied Japan after the Pacific war. Secondly, the Americans have had the capability—specifically the naval and air power—to protect the Japanese islands against Soviet attack, while the Soviets, powerful as they have been on the Eurasian continent, have not had the means to protect Japan against the United States. Thirdly, American military power can not only protect Japan itself, but can also insure Japanese access to the world oceans and to the distant sources of energy and raw materials necessary to Japan's viability as an industrial economy. The conservatives also abhorred Soviet communism; they preferred parliamentary, constitutional government, and they were impressed by the good will displayed by U.S. Occupation policy. But for purposes of reflecting on strategic military problems, it is sufficient to note that all of Japan's prime ministers have believed that Japan is not able to defend itself against either of the military superpowers, that the Soviet Union cannot defend Japan against the United States, but that the United States can defend Japan against the Soviet Union. Accordingly, Japan has kept its own forces small and allied itself with the United States in the 1951 security treaty and, since 1960, under the terms of the Treaty of Mutual Cooperation and Security.

The question of what kind of alliance relationship the Japanese have wanted to build is also understandable in simple, straightforward language. They have wanted the United States to guarantee Japan's external defense and to maintain sufficient forces in the Far East to make that guarantee into a credible deterrent. They have not wanted the U.S.-Japanese alliance to pose an offensive threat to the Soviet Union or communist China, a threat which, to their minds, would increase the possibility of a major war in which Japan would be a battleground. The friction that has existed between Tokyo and Washington over the security treaty is largely understandable, therefore, as a product of the differences between Japanese and American estimates of the threats to Japanese security and the measures necessary to defend Japan.

The two main issues in this regard have been the size and equipment of the U.S. forces in Japan and the rebuilding of Japan's own armed forces. In general, the U.S. government has tended to make a higher estimate first of the Soviet and then of the Chinese threats. Washington has assumed that Japan's security is inseparable from the security of the Far East. It has sought to use Japan as a base for Far Eastern operations, from the Bering Straits to South Vietnam. And Washington has argued the need for powerful Japanese armed forces. The Japanese government has made a lower estimate of the Soviet threat and a much, much lower estimate of the Chinese communist threat. They have believed that the U.S. ground forces in South Korea, the American naval and air forces in the western Pacific, and American nuclear missiles are a credible and effective deterrent against any hostile military action in the Japan area. They have favored sufficient U.S. bases in Japan to maintain this disposition of forces, and they have been willing to build the small, conventionally armed Self-Defense Forces. These forces have been intended mainly to raise the threshold of a Soviet attack against Japan, thus strengthening the American guarantee.

Neither the publication by the Defense Agency of its first White Paper in the fall of 1970, nor the planned doubling of defense expenditures under the Fourth Defense Build-up Plan (1972-1976), suggests that the Japanese government intends to change this policy or to fundamentally alter the role of the Self-Defense Forces. The White Paper says nothing new. It stresses that it is the duty of the Self-Defense

Forces to protect Japan against limited, conventional attack. It argues for greater "defense consciousness" (*boei ishiki*) on the part of the Japanese people, and stresses the need for more self-reliance and less dependence on the American guarantee. This is what conservative prime ministers and dietmen have been saying since the mid-1950's.

Upon examination, it appears that the projected doubling of defense expenditures under the Fourth Plan will not significantly alter the role or the relative capability of the Self-Defense Forces. The single largest item in the plan will be the procurement of 170 F4J Phantom jets, to replace the Air Self Defense Force's aging F104s. The second largest item will be the replacement of a number of obsolete naval vessels, and the building of additional tonnage, including one or two new destroyers armed with missiles, and a helicopter carrier for anti-submarine operations. The rest of the budget increases will be consumed by pay increases and refurbishing of Self-Defense Force facilities, many of which have been getting less than minimal maintenance as U.S. forces have evacuated them. In view of continuing increases in Soviet and Chinese military capabilities, both nuclear and conventional, Japan's *relative* military strength in 1976 is not likely to be greater than it was in 1966. Finally, there is no evidence in the Fourth Plan that the Self-Defense Forces will increase their ammunition reserves, which have throughout the 1960's been less than their thirty-day requirement. Unless these ammunition reserves, and fuel reserves as well, are dramatically enlarged, Japan will continue to rely on United States air and naval forces for defense against large-scale or protracted conventional attacks. In short, under the Fourth Plan the mission of the Self-Defense Forces will continue to be limited to maintaining internal security, and to repelling low-level conventional attacks and probing actions.

Although the available evidence indicates that as the 1970's began the Japanese government was not planning a new defense policy for the new decade, it is possible that changes in Japan's domestic and international environments will lead to new departures.

Domestically, there seem to be two broad possibilities. First, the conservatives may suffer a schism over the basic premises upon which their policy has been based. If we look at Japan's spectacular economic growth, at the increasing sense of national pride among the Japanese, and the Chinese nuclear program, it would appear that the premise over which the LDP is most likely to split is the belief that Japan is not able to defend itself against either of the superpowers. Hatoyama's abortive policy of rapprochement with the Soviets, accelerated rearmament, and greater independence toward the United States still has its champions in the LDP. Discussions and debates among conservatives over the NPT suggest that the central issue here is whether Japan should, *in the future*, become a nuclear power. For the time being Prime Minister Sato has established an agreement in his party that enables Japan to adhere to the NPT. It seems that this agreement is based on the understanding that Japan will continue its peaceful nuclear development and its rocket testing program. The NPT is not interpreted as an absolute prohibition against building nuclear weapons. It is rather a convention representing an international consensus against proliferation. If, in the future, any of the non-nuclear states, including Japan, believes that its basic security is jeopardized by the convention, it can withdraw and go nuclear.

A second domestic change that could radically alter defense policy is a collapse of the LDP followed by the formation of a coalition government headed by socialists. Although the socialists have been calling for unarmed neutrality since 1949, they would find it extremely difficult to translate their slogan into a policy. To begin with, they would probably have to accept the existence of the Self-Defense Forces. Although the Self-Defense Forces have relatively little interest or influence in politics, most socialists believe that an order to disband or to convert into a disaster relief corps might provoke an attempted military coup. In all likelihood, the socialists would settle for cuts in the defense budget and a reaffirmation of civilian control of the Defense Agency and the Self-Defense Forces. The socialists might very well terminate the security treaty, however, and attempt to drum up domestic support by declaring that Japan is "neutral" and has an "independent foreign policy." If the United States, the Soviet Union, and China respected that neutrality, this might prove temporarily feasible. If, however, this neutrality were not respected and the socialists perceived military threats to Japan, they or their successors would once again either have to team up with one of the superpowers, go nuclear, or do both. It may seem paradoxical to contend that the pacifist, reformist socialists could make Japan a nuclear power. However, they are strongly nationalistic; they want independence from the Americans; they suffer from the same contradiction between humanitarian reformism and emotional nationalism that plagued the pre-World War II Japanese liberals. Once in power, their reformism, like that of their predecessors, might fall victim to their nationalism.

Assuming that the conservative leadership does hold together and does stay in power through the 1970's, and that it wants to stick to its defense policy, it might nevertheless still be forced to make a new policy as a consequence of fundamental changes in Japan's international environment. In speculating on external changes, it is possible to set off on an endless guessing game predicated on possible shifts in U.S., Soviet, Chinese, North Korean, and South Korean policy, and the effects major departures in the policy of these states would have on the others, including Japan. But this guessing game is not necessary. For it is extremely difficult to conceive of any moves by Japan's Asian neighbors that would undermine Japan's defense so long as Japan has a firm, reliable security relationship with the United States. If the conservatives stay in power and want to continue their present defense policy, they would be forced to alter it only if one major change occurs in the international environment. That change would be the perceived loss of the capability or the willingness of the United States to defend Japan. If, in the aftermath of the Vietnam War, the United States drastically reduces its military presence in Asia and the Pacific Ocean, or unilaterally and without Japanese cooperation undertakes to restructure its relations with China or the Soviet Union, or pursues a protectionist foreign trade policy that seriously damages Japan's economy, then the conservative leadership will be compelled to either devise a new defense policy, or to make way for fresh leaders who will do so.

Thus, the most likely alternative is the one most discussed by foreign observers: accelerated rearmament, including nuclear weapons, accompanied by a weakening of the alliance with the United States and directed at achieving the highest possible level of military self-sufficiency in the shortest possible time. But in the military as in the economic field, the concepts of the twenty-year-old policy of balanced defense have become so ingrained in conservative thinking that only a major shift in political leadership in Japan or a fundamental change in relations with the United States is likely to lead to the abandonment of the present policy.

CHAPTER III

The Stability of Conservative Party Leadership

NATHANIEL B. THAYER

SECURITY POLICY like other major policy in Japan is the product of interaction among the business federations, the bureaucracy, and the conservative politicians, with the businessmen usually insisting on what is profitable, the bureaucracy on what is rational and economical, and the politicians on what enhances themselves and the nation.

And what of the wishes of the people? Japan is after all a parliamentary democracy. The constitution gives the people ultimate authority: they elect the legislators, the legislators elect the prime minister, and the prime minister appoints the cabinet. But Japanese voting behavior is so structured and the election law is so designed that most politicians find they must create and run on their own policies. These personal policies rarely become either party or national policy.

In the national legislature (Diet), debate is jejune. It is usually given over to extravagant attacks by the opposition politicians, with the conservative politicians saying as little as possible to avoid giving ammunition for further attacks. If there is a focus for national policy, it is the cabinet. The *Asahi shimbun*, one of Japan's largest national dailies, has been making surveys at least twice yearly for the past fifteen years, comparing popular support of the cabinets with that of the conservative party that produces them. These surveys have shown that support for the conservative party is high and consistent, but that support for any given cabinet is low and volatile. The prime minister changes the cabinet regularly, at least once a year, sometimes more often. He accomplishes three things: first, he is able to avail himself of whatever particular expertise he needs at the moment; second, he is able to consolidate support within the party by an astute reassignment of high posts; third, and sometimes most important, he is able to use the dismissed cabinet as a safety valve, making its last function to absorb public dissatisfaction.

There is a further safety valve. That is the replacement of the prime minister. When his popularity falls too low, the other faction leaders in the party begin to conspire against him; however, the conservative party's tolerance for unpopularity is great. Prime Minister Hatoyama's rating in the public opinion polls fell below 30 percent and Prime Minister Kishi's to just above 10 percent before their colleagues replaced them. Even in these cases the replacement of the premier did not mean the reversal of major policy. One may safely conclude that unless there is an overwhelming shift in national sentiment, the conservatives are under no compulsion to shape the nation's security policy in response to public emotions. So long as a working arrangement can be sustained with the top bureaucrats and big businessmen, the conservative political leaders will feel free to adopt the security policy they believe is right. The future of Japan's policy of balanced defense depends, therefore, on the continuance in power of Japan's conservative party leadership, which in turn depends on the stability of the political system that has evolved over the postwar years.

The first question to be raised about the future, then, is: how stable is this political system? Is it reasonable to suppose that it will remain relatively unchanged during the next five years? Four potential threats to the Japanese political system need to be considered: external attack, internal coup or revolution, revision of the constitution, and reform of the electoral law.

The likelihood of external attack on Japan seems remote, and the chances of the success of such an attack even more remote. The European powers previously involved in Asia are withdrawing. The developed nations of Asia have profitable trade ties that war would only disrupt. The underdeveloped Asian nations receive Japanese aid; furthermore, their armies could not mount a sustained overseas attack. Most non-communist Asian nations have military agreements with the United States and rely heavily on it for logistic support. The United States would probably be able to dissuade them from military action against Japan.

The communist nations of Asia pose a slightly greater risk to Japan. North Vietnam can be ruled out as a challenge; the distance from Japan is too great. North Korea might choose to harass Japan militarily, but it could not hope to win an all-out attack. Russia and China should be deterred by the guarantees given the Japanese under the U.S. security treaty. In sum, Japan has little to fear from the outside.

Another possibility is overthrow of the present government by internal forces, either through a military coup d'état or a popular revolt. The left-wing Japanese parties constantly charge Japanese military forces with preparing for a coup d'état. In the light of Japan's history, it is a legitimate concern. Yet it is difficult to equate the present military forces with prewar military forces. The fear of subjection that gave rise to the preeminent position of the military forces in prewar society no longer exists—at least for the Japanese. The nineteenth-century desire for empire has been sublimated into a twentieth-century desire for trade. The prewar constitution that gave the military special powers has been replaced by a new constitution that relegates the military to a subordinate role. Most importantly, the prewar military officers perceived their role differently from the postwar officers. The former regarded themselves as chosen to lead the Japanese nation; the latter clearly recognize civilian supremacy. It is difficult to accept the thesis that the Japanese military forces still constitute a threat to the state.

Popular revolt is another alternative, either through a mass uprising or through revolutionaries acting in the name of the public, but the idea of a mass uprising is incredible. There are no symptoms of disease or even debility in the Japanese body politic. Although the Japanese citizen usually can find something to complain about, public opinion surveys show him to be basically satisfied. The possibility of revolutionaries seizing power is only slightly less incredible. Like any other state Japan has its political fringes. From time to time the public media uncover a plot among unreconstructed members of the far right. Japan has its admirers of Mao, who subscribe at least theoretically to violent overthrow. Anarchism has many student adherents. These individuals and groups, however, constitute only a thin line on the political spectrum. They enjoy little popular sympathy and no popular support. They represent a police, not a political, problem.

Constitutional reform would also significantly affect the Japanese political system. Formally, Japan has had two constitutions, though some historians regard old military codes and the house laws of the various feudal estates as constitutions as well. The first formal constitution was written in the late nineteenth century and was strongly influenced by German constitutional thought. This constitution, the Meiji constitution, was replaced in 1947 by the present constitution, which was written during the Occupation and reflected American ideas. When Japan regained its independence, some Japanese urged that it be rewritten.

The advocates of constitutional reform are generally conservatives, many of whom occupy important posts in the government. They argue that the document is too liberal, that the constitution of the country should be framed by Japanese and reflect Japanese concepts and traditions. The minority parties, on the other hand, generally uphold the present constitution. They agree that it is a liberal document and for that reason urge its retention. They support Article 9, which prohibits Japan from maintaining a war-making potential. They argue that any substitute constitutions written by the conservatives would be "feudal."

Since the conservatives control the government, it would seem that constitutional reform would be a real possibility, but such is not the case. Technically, the conservatives would have to muster a two-thirds majority to vote amendments or a new constitution through the Diet, and they do not have this number of seats. Furthermore, the conservative party is far from united in the belief that the constitution should be rewritten. Finally, surveys show that the public is generally satisfied with the present constitution.

Although constitutional reform is still debated, particularly around election time, the issue is dead. To satisfy the party's right wing, each prime minister states that he will study the problem, but does nothing more. This is not to say that the constitution will not change. Like any other document, it is capable of varying interpretations, though in Japan probably the national legislature and the bureaucrats rather than the judiciary will find new meanings. But changes will be gradual, certainly will not challenge the present ruling party, and will not materially affect the leadership of Japan in the short run.

Electoral reform is another measure that would change the Japanese political system. The supreme political authority in Japan is the Diet, which is composed of the House of Councillors and the House of Representatives. The two houses, however, are not equal. Traditionally, the lower house, the House of Representatives, is more active. The constitution makes it superior to the upper house, the House of Councillors, by giving it special powers in treaty-making and budget formulation. The electoral systems for the upper house and the lower house differ. Recommendations for changing the election system for the upper house have been debated from time to time. Changes would greatly affect the fortunes of the councillors but would not materially change the destiny of the nation.

Changes in the electoral system for the lower house could have profound affect. The electoral system for the lower house was conceived in 1925. It was based on two principles: first, that the electoral seats should be divided equally among the population and, second, that existing political divisions should be maintained. At the time there were 466 seats in the lower house. This figure was divided into the total number of enfranchised voters to determine how many voters there should be for each representative.

Then as now, Japan had twenty-seven prefectures. The figure representing the ideal number of voters was divided into the population of each prefecture. Fukui and Tottori prefectures were found to have sufficient population for four representatives and these two prefectures became four-man election districts. Yamanashi and six other prefectures had populations sufficient for five representatives and they became five-man election districts. The rest of the prefectures had larger populations. They were subdivided so that no electoral district would have more than five representatives. Aomori, for example, had sufficient population for seven representatives and was split into one four-man district and one three-man district; Osaka was split into five electoral districts; Tokyo was split into seven. The voter has one ballot, which he casts for one candidate. Candidates receiving the greatest number of ballots are sent to the Diet. Sixteen elections have been held under this system.

The system seems logical enough, but it has produced illogical results. The Japanese voter is a remarkably constant fellow. He rarely switches his vote from party to party, though he may switch his vote around among the candidates of one of the parties. The election system, then, encourages political fighting, not between the candidates of the different parties, but among the candidates of the same party. The minority parties have generally solved this problem by limiting their candidates. In an election district where they have only enough votes to put one man in office; they run only one candidate. If they have enough votes for two candidates, they run two candidates. The conservative party also tries to limit its candidates, but is far less successful, because many conservative candidates will run as "independents" if they do not receive the party endorsement, with the result that there are far too many conservative candidates in each election.

The conservatives complain bitterly about the electoral system. It allows minority candidates to sneak in and take over seats when too many conservative candidates fragment the vote. It encourages dissension and precludes cooperation within the party. It makes party platforms meaningless. Substantial political monies are spent fighting not other party candidates but candidates of the same political persuasion. Instead of bringing about the emergence of a two-party system, which Japanese political commentators admire and wish to emulate, the present electoral system encourages the continuance of a multi-party system.

An electoral commission has been studying the problem for many years and various possibilities have been reviewed. The commission has concluded that the present districts should be made into single-member districts, that is, a district in which only one seat for the Diet would be at stake. Although debate continues, the newspapers and the academicians have accepted the commission's recommendations. The Liberal Democratic party, the ruling party, has stated that it would work for electoral reform.

Nevertheless, chances for the passage of a new electoral law are slim. First, all the opposition parties agree that they would lose seats under a single-member district system and are therefore adamantly opposed to it. Second, although the Liberal Democratic party has said that it favors single-member districts, not all members of the party are in accord since many of them would be forced from office. Third, within the LDP are factions, and these factions gain members in large measure by assisting conservative candidates in the general election. Their leaders would resist any reforms that would strengthen the central party at their expense. For the short run at least, the Japanese political system can be expected to remain pretty much as it is.

It would be possible, of course, even operating within this stable system, for new political forces to emerge that could affect the leadership and its policies. To have a serious impact during the first half of the 1970's, however, such forces presumably would need to be already organized. Major organizations worth considering, therefore, are lobbies, the students, religious groups, labor unions, business federations, and the local governments.

By definition, lobbies are interested in only one part of the body politic although sometimes one does expand its interest to areas beyond the reason for its creation. They are relatively new to Japan, having come forth during the postwar years, if not the last decade. Initially they were subjected to harsh criticism, but more recently the politicians have come to realize that the lobbies can benefit the nation even though they are pledged to work for special interests. In the past, the farm and fishing cooperatives and the veterans' associations have been the most powerful since they had simple demands and a large membership distributed nationally. Today new lobbies are emerging and many of them are working at cross purposes. The institution of the lobby is coming to be accepted in Japan and can be expected to grow in importance. But there appears to be a balancing of influence among them, so that no single lobby or group of interests is able to dominate the political scene.

Overseas groups form a distinct category. They include not only the Japanese emigrants, such as the Brazilian farmers, but also the Japanese business and trade organizations that have started overseas operations. In the past the former groups have not mattered too much, although Japanese concepts of nationality have always given them full standing within the Japanese polity. Japanese international organizations are growing in importance, and their voices will be louder and heard more often. So far, however, they have been willing to have their interests represented by the traditional business federations. No doubt they will eventually demand individual representation, but this development will probably not occur during the early 1970's.

Students deserve more careful consideration, especially when one recalls the extraordinary mischief that the young officers were able to create in the days before the war. In the postwar world students have played a highly visible role, invariably opposing authority and expressing their opposition actively. In the 1967-1969 period, for example, they were able to paralyze the Japanese university system and precipitate a national crisis. But excesses have turned the public against the students. Even at the height of their power, students were able only to disrupt the campuses and jam traffic in the streets. Now, in their weakened condition, it is hard to imagine them having the muscle and imagination to do even that.

A noteworthy phenomenon in postwar Japan has been the extraordinary growth of the new religions. The most successful has been the Soka Gakkai (the Value Creation Society). Utilizing modern techniques of organization, it has secured many converts, and its political arm, the Komeito or Clean Government party, has become an important force in the Diet. The established sects, which many observers considered moribund, have emulated the new sects, at least in their proselytizing skills, and have been able to gain a new purchase on the Japanese citizenry. The religious bodies, then, should be scrutinized in looking for new sources of political power.

The Soka Gakkai is in a league by itself. In 1960 its leaders decided to enter politics and in just a decade their political party has become the third largest in the Diet. Although political commentators have freely predicted its decline and demise, in each election it has grown. This extraordinary growth may be attributed partly to its skill in taking advantage of the anomalies in the election law. It has run candidates in large urban election districts where it has had a big enough cluster of votes to elect a candidate, and in the prefectures constituting a single election district. No doubt it will continue to grow, particularly since the public has become increasingly disenchanted with other minority parties. But it is unlikely that the Komeito will be able to become the majority party. To do so it must broaden its appeal to pick up a greater number of voters in a greater number of districts. Specifically, the Komeito must attract voters other than believers in the Soka Gakkai and these voters must come from the rural areas. So far it has had only modest success in this venture.

The other new religions have also demonstrated political ambitions. As yet, none of them has been able to place members in the lower house, although they have elected members to the upper house. For many of these new religions membership in the upper house is sufficient since they are searching for prestige, not political power. Conceivably some of them may be able to accumulate a large enough bloc of votes within one of the major cities to elect a candidate to the lower house. But at the present writing none shows that potentiality. In theory, the new religions (other than the Soka Gakkai) might unite to run a single slate, but each has its own world (or other world) view antithetical to that of the others. Religious intolerance makes political cooperation difficult. The older established religions have exhibited only a slight political thirst, which has been slaked by working through the established parties, most generally through the Liberal Democratic party.

Labor unions can be passed over quickly. Each of the labor federations has identified itself with one of the minority parties, thus making its fate dependent on the fate of the party. The Japanese electorate has not as yet shown a desire to allow any of these minority parties to rule. Conceivably, new unions may appear on the scene or the old unions may cast off their policy of working through a particular party. Conceivably, such unions may prove attractive to the voter. But it is unlikely for this process to take less than a decade.

Business federations also do not need lengthy consideration. They have completely identified themselves with the Liberal Democrats. The business federations have given the party the funds it needs to operate, and the party has given the business federations a good environment in which business can flourish.

Local governments do not wield any significant influence at present. Since the seventeenth century, political power has emanated from Tokyo. Local governments have had no greater role than to carry out the central government's orders and deliberate matters in which the central government has shown no interest. This situation may be changing. Japan is becoming more difficult to rule as its society becomes more complex. Like citizens in other industrial countries, the Japanese often complain that the central government is impervious to the wishes of its citizens. As in the United States, voices call out for more power to local government. But the United States has a tradition of local rule; Japan has not. Political power may eventually devolve to the local Japanese bodies. In such a case, the local assemblies will become testing grounds for the national Diet. But such a move is light years away. Today there is a strong trend to draw new politicians from the central bureaucracy.

Consequently, it is exceedingly unlikely in the foreseeable future that political forces in Japan will change. Our next task is to investigate whether the balance of power among the existing political parties will shift.

The general election decides which party will rule Japan.

Under the new constitution the emperor still has the formal power to dissolve the House of Representatives and proclaim general elections, but he must act on the advice and approval of the cabinet (Art. 8), the majority of whose members must be chosen from the Diet by the prime minister. In short, the power to call an election devolves upon the prime minister. The only stricture the constitution places on him is that he must call a general election at least once every four years. The years in which general elections have been held since the end of World War II are given in Table 1.

TABLE 1

YEARS IN WHICH GENERAL ELECTIONS HAVE BEEN HELD IN JAPAN[1]

1946	1953	1963
1947	1955	1967
1949	1958	1970
1952	1960	

[1] The source for these and all subsequent election statistics is the official publication *Kokka shirabe* (National Survey), issued by the Home Ministry after each election.

During the early part of the postwar period elections were held relatively frequently, on the average of once every two plus years. Since 1960, however, prime ministers have been content to let the Diet run almost to its constitutional limit. If events continue in their present course, only one—at the most two—elections can be anticipated in the period 1970-1975.

The last general election was held in 1970. The results are given in Table 2.

A short analysis of those figures demonstrates that the Liberal Democrats hold a majority of the seats in the lower house. Whether or not they hold a majority of the vote is not clear. As explained above, the election law causes many conservatives to run as independents since all cannot secure party

TABLE 2

RESULTS OF THE 1970 GENERAL ELECTION IN JAPAN

	No. of Seats	*Percentage of Vote*
Liberal Democratic party	288	47.6%
Japan Socialist party	90	21.5
Komeito	47	10.9
Japan Democratic Socialist party	31	7.7
Japan Communist party	14	6.8
Other parties	0	.2
Unaffiliated	16	5.3
Total number of seats	486	100%

endorsement. The most recent election was no exception. Of the sixteen unaffiliated candidates who won, twelve chose the day after the election to "join" the Liberal Democratic party, thus giving to the conservatives three hundred seats. The originally unaffiliated candidates accounted for 5.3 percent of the vote: part of this vote belonged to the Liberal Democrats, perhaps enough to put them over the 50 percent mark. The rest of the vote was fragmented. The Socialists, who comprised the next strongest party, were only able to win ninety seats or 21.5 percent of the vote.

Drawing conclusions from one election is risky. But this election, when placed in historical context with other elections, demonstrates certain trends in the conservative party. Table 3 gives the results for the conservatives of all general elections since 1953.

The conservatives have held either a majority or a high plurality of the vote since 1953. Over the years, however, the vote has gradually decreased, although the rate of decrease has declined in the past two elections. This drop in the percentage of the vote has not been translated directly into an equal drop in the percentage of Diet seats held by the party. The last election shows, in fact, a drop in the percentage of the popular vote but an increase in the percentage of seats

TABLE 3

HISTORICAL ANALYSIS OF THE RESULTS FOR THE CONSERVATIVE PARTY OF THE GENERAL ELECTIONS SINCE 1953

Year	*Percentage of the Vote*	*Percentage of Number of Seats*
1953	65.7%	66.5%
1955	63.2	63.6
1958	57.8	61.5
1960	57.6	63.5
1963	54.7	60.6
1967	48.8	57.0
1970	47.6	59.3

held. If the analyst does not insist on too great a degree of precision, the results of the next two elections are predictable. The conservatives will continue to lose votes although they will still maintain a high plurality. This loss, however, will probably not mean that the conservatives will lose their absolute majority in the House of Representatives.

The second largest party is the Japan Socialist party. The height of its strength was reached in 1958 when it occupied 166 of 467 seats in the lower house. In the four elections since that time, its strength has declined. The last two elections were particularly disastrous. In the 1967 elections the socialists were able to win only 144 of 486 seats; in the 1970 elections the number of seats dropped precipitously to 90. For better than twelve years the Socialist party has been in steady decline. The only party which shows vigorous growth is Komeito, and as yet it has been able to win only 47 seats. Little chance would seem to exist for any single minority party to take over the government in the near future.

The possibility is slightly greater for the conservatives to lose an absolute majority of seats in the lower house, permitting, at least in theory, a coalition of the minority parties to take over the government. But this too is not very likely to happen. Even if the conservatives were to lose their absolute majority, they would probably not lose it by much. Moreover, a coalition would have to include all the minority parties. While these parties have been able in the past to find common ground on which to oppose the conservatives, the ideological distance between them is so vast that it is difficult to imagine their being able to agree on a comprehensive program of rule. It is impossible to imagine them being able to agree on a single man to serve as prime minister.

A variant of this possibility would be for the present conservative party to splinter, permitting some conservative elements to unite with one or more of the minority parties or parts of the minority parties to form a new party or a new coalition. To evaluate this possibility one needs to know something of the political behavior of rural Japan.

Inasmuch as the electoral system forces conservative to fight conservative, the party can play no role in these fights. It must treat all its candidates equally. Gathering votes, therefore, becomes the individual responsibility of each politician. In many instances he does not deal with the voter directly but works indirectly through vote brokers.

The Japanese voter long ago learned that strength lies in numbers. This sentiment is particularly strong in small rural villages, which tend to vote as units. It is not unusual for a village to give as much as 85-90 percent of its vote to a single candidate. The village headman, or other distinguished elder, decides who will get the village's vote, usually on the basis of who will promise and deliver the most for the village.

These village vote brokers almost always deliver their vote to the Liberal Democratic party, not because it is conservative, but because it is the party in power and the greatest largesse flows from the government. While loyal to the conservative party, many of these brokers feel free to shop around among the individual conservative candidates. An examination of the statistics of many of the election districts shows that the number of votes for the conservatives remains much the same, but that the number of votes for each of the conservative candidates in the district fluctuates greatly.

A revolt or a threat of revolt among the conservatives can be handled easily by the prime minister (who heads the conservative party). All he need do is hold elections. Unless a revolting conservative can demonstrate convincingly that he and his cohorts are stronger than the prime minister—and generally the dissenter cannot—the vote brokers generally will desert him and give the village votes to another conservative who has the support of the prime minister.

To be sure, not all election districts are of this traditional, rural type. In recent years the cities have grown immensely, and they have their own distinct political behavior. Furthermore, changes have also occurred in the countryside. Today many politicians make direct appeals to the electorate and get elected. But still, enough of the traditional pattern of behavior exists in enough of the election districts to make a revolt within the conservative party unlikely.

Thus, the probability that a new conservative party will emerge is dim. For the immediate future, Japanese leadership will continue to be drawn from the present conservative party. The crucial question then becomes: who will lead that party during the next five years? The answer is surprisingly predictable due to the factional nature of the party.[1]

The present conservative party was created in 1955. In most ways it resembles the earlier conservative parties, but in one important way it differs. The Diet now holds the power to select the prime minister. Article 67 of the constitution states: "The prime minister shall be designated from among the members of the Diet by a resolution of the Diet." This means that as long as the new conservative party is able to

[1] For a more extended discussion of the functions and structure of the LDP, see Nathaniel B. Thayer, *How the Conservatives Rule Japan* (Princeton: Princeton University Press, 1969).

hold onto a majority of the seats in the lower house, its president serves as prime minister. Under party law the president is elected for a term of three years. To become president, a candidate must obtain a majority of the votes cast by every party member in the Diet and a delegate from each of the forty-six prefectures. In the event no candidate receives a majority on the first ballot, a run-off election is held between the two candidates receiving the greatest number of votes.

The rules for the election of the party president were written without much study when the party was created. No one was quite sure what they really meant. Their significance became apparent as the party strongmen began jockeying for power. First, because the franchise was limited, hovering between four and five hundred votes, a strongman could approach each voter directly and ask for his support. Second, the voters were all politicians; thus, appeals were political. Third, the presidential and the popular elections were out of phase, and each dietman would probably vote in several presidential elections. Fourth, since coalitions would have to be formed to obtain a majority, a strongman would have to be able to swing the votes of his supporters from candidate to candidate if he were to participate meaningfully in the electoral process. The strongmen soon realized that they needed permanent rather than temporary supporters. They began organizing clubs or factions.

To get, the strongman had to give. Although the party could not favor one candidate over another in fights between conservatives in the general elections, the faction leaders certainly could. They offered experience to new candidates; they gave money to poor candidates; they sent popular speakers to whip up enthusiasm for unknown candidates. These services proved invaluable. A strongman's support became almost indispensable to winning an election.

The party presidential elections further strengthened the

strongmen. Initially there were better than half a dozen strongmen. Success in winning a presidential election meant success in putting together a coalition. Front runners soon found that they could secure the support of other strongmen by promising them posts, particularly cabinet posts, which they could distribute to their followers. Not too many years passed before it became impossible to occupy a cabinet seat without the recommendation of a faction leader. Dietmen who aspired to a cabinet post joined factions.

As each party election and popular election took place, the ties between the leader and his followers became firmer. Today the factions are an integral part of the conservative party, with faction leaders the job brokers for their followers and themselves the exclusive contenders for the premiership. In theory, of course, the faction leaders could choose for the highest office another dietman. (The constitution specifies only that the prime minister must come from the Diet.) But they have not done so in the past, and there is no apparent reason why they should do so in the future. A forecast of the quality of leadership of the LDP in the 1970's, therefore, can be concentrated on the characteristics of its faction leaders.

It is important to point out at the outset that these are not the kind of men one might expect to rise to the top in a Western parliamentary regime. Activity on the floor of the Diet or in its committees is not one of the qualities needed by a faction leader. When Prime Minister Sato took office, the reporters combed the Diet records for his speeches, hoping to find clues to his policy. Although Sato had served seventeen years in the Diet, they found only one speech and that was a eulogy for another politician.

Nor is having charisma important although several faction leaders do. The ability to write convincingly, to speak well, to make an impressive public appearance—the public attributes we associate with Western politicians—are not needed. The making of a faction leader takes place among a small group of men within the Diet. All a prospective faction leader must do publicly is to keep winning elections in his home district, and many times he can do that through negotiation with the vote brokers.

The faction leader need not be known as a policy maker. This is not to say that the faction leaders have no ideas. Ikeda, for example, had been a bureaucrat in the Finance Ministry, and his faction was always concerned with economic questions. Kono served often as agriculture minister, and his faction followed rural affairs quite closely. Miki was educated in the United States and has always been interested in foreign policy. But these are predilections.

The factions have always been whipping boys for the public media, which have made a major issue out of this lack of interest with policy. Their outcry has had some effect on the faction leaders. During early party presidential elections no faction leader felt obliged to enunciate his policies, even though he was in a contest for the highest office in the land. Nowadays, however, each serious contender for the premiership feels obliged to issue position papers and give a statement of his views. The policy papers are usually written by some bureaucrat, and the statement of views by the faction leader's secretary. All statements are general. If they antagonize no one, it is because they do not say very much.

It has been argued that too close an identification with policy works to the detriment of the faction leader. A clear position on a policy may win plaudits from some people, but it makes enemies of others. The most important virtue of a faction leader is to be all things to all men.

The LDP factions are quite limited in number. Originally there were seven; now there are eleven. The twelve leaders (one faction has dual leadership) cannot be expected to monopolize the control of the party until 1975, for succession problems within several of the factions cannot be avoided. Retirement from the premiership, for example, has usually led to retirement also from factional leadership. This will probably eliminate Eisaku Sato sometime within the next year or so. Natural causes or voluntary retirement also can be expected to take their toll since four leaders (Fujiyama, Funada, Ishii, and Shiina) are over seventy years of age.

It is not possible to predict how succession problems will be resolved, since one of the peculiarities of faction organization is that there is no hierarchy. Although some men in the faction seem closer to the faction leader than others, they cannot properly be regarded as lieutenants, for a faction member's relation to the faction leader is direct: no one stands between them. Thus, when the faction leader dies or retires from his responsibilities, he has no readily accepted successor. The faction often splits under rivals within the old faction; it may disappear in whole or in part as the members get recruited into other factions, or it may hang together around a new champion drawn from within its midst.

As summarized in Table 4, when Prime Minister Kishi stepped down, for example, he realized that the odds were greatly against his serving as prime minister again. He was, however, determined to preserve his political power and choose a future prime minister. He chose a successor, Takeo Fukuda, and over the succeeding years has worked to transfer the faction's loyalty to him. Fukuda has proved strong enough to hold a large part of the faction together, but Kawashima and Fujiyama each split off to form his own successor factions as well.

Another original faction leader, Ichiro Kono, died in 1964. Yasuhiro Nakasone, one of his followers, attempted to take over leadership but was rebuffed. After various attempts at group and committee leadership, the faction split, with Nakasone becoming the leader of one of the splinter factions. Through the 1967 and 1970 elections Nakasone was able to bring many of his own candidates into the Diet and through diligence and political skill was able to win back many though not all of the Kono followers who went off with the other factions. Today Nakasone is regarded as Kono's successor, although in truth he had to build his faction almost anew.

When Hayato Ikeda became prime minister in 1960, he appointed a directorate to run his faction. On his death in 1964, it was natural for this group to assume power. They passed the leadership on to Maeo, a close confederate and companion of Ikeda, but this was an interim appointment until another leader should emerge. Today the newspapers regard Masayoshi Ohira as the successor, though there are still men in the faction who dispute the newspapers' judgment. No doubt Ohira will have to follow a course of consolidation over the next few years.

Finally, Kakuei Tanaka is trying to inherit the Sato faction. This is delicate business. Eisaku Sato is still prime minister, the most powerful politician in the country, and he has given no signs of a willingness to allow his faction to drift apart. Nevertheless, Sato will step down some day and Tanaka wants to be ready when he does. The newspapers declare his search for followers to be successful, although he too will not be able to carry away the whole faction. Tanaka's task appears to be similar to Nakasone's and a little more difficult than Ohira's. (Some members will choose to ally with Fukuda, the successor to the Kishi position. Kishi and Sato are brothers.)

While the LDP factions can be expected to change substantially during the next five years, the number of potential premiers is limited. First of all, the economics of factions must be considered. The ideal size of a faction seems to be about twenty-five lower house politicians. Except for the prime minister in power, other faction leaders can rarely muster the funds to support more politicians. Fewer members tend to downgrade the strength of the faction. Even the most euphoric party booster cannot conceive of the conservatives winning 300 seats in the next election or the one after that. Simple arithmetic suggests, then, that the maximum number

TABLE 4

LDP FACTION LEADERS, 1955-1975

Original Leaders, 1955	*Initial Successors*	*Present Leaders, 1971*	*Probable Leaders to 1975*
RULING FACTIONS (HAVE HELD PREMIERSHIP)			
Eisaku Sato (69, premier since 1964)	———	(will probably split when Sato retires from premiership)	KAKUEI TANAKA (52)
Nobusuke Kishi (74, ret. from 1960)	TAKEO FUKUDA (66)	———	———
	Aiichiro Fujiyama (73)	———	?
	Masajiro Kawashima (d. 1970)	Etsusaburo Shiina (73)	?
Hayato Ikeda (premier, 1960-1964, d. 1964)	Shigesaburo Maeo (65)	MASAYOSHI OHIRA (60)	———
OPPOSITION OR SUPPORTING FACTIONS (HAVE NEVER HELD PREMIERSHIP)			
Ichiro Kono (d. 1964)	Collective:		
	Kiyoshi Mori (d. 1968)		
	YASUHIRO NAKASONE (52)	———	———
	Sunao Sonoda (57)	———	———
	Munenori Shigemasa (defeated 1969)		
Matsumura-Miki	TAKEO MIKI (63)	———	———
	Kenzo Matsumura (d. 1971)		
Mitsujiro Ishii (81)	———	———	?
Bamboku Ono (d. 1965)	Isamu Murakami (68)	Mizuta (Mikio, 64)-Murakami	———
	Naka Funada (75)	———	?

NOTE: Age as of January 1971 is given in parentheses. Names of those who are reali possibilities as successor to Sato in the premiership, 1972-1975, are given in capital letters.

of factions will be twelve. A more pessimistic estimate might place the number of seats around 270 and the number of factions about ten. Although one can be sure that the prime minister will be a faction leader, not all faction leaders hope to become prime minister. Some of the faction leaders are now too old. It seems to be an unwritten law that seventy years of age is the cutoff date for consideration for this office. Others have lost their ambition. Still others have problems of health. For these reasons, of the actual and potential faction leaders today, there appear to be only five who can be considered for the office: Takeo Miki, Yasuhiro Nakasone, Takeo Fukuda, Masayoshi Ohira, and Kakuei Tanaka.

TABLE 5

POLITICAL EXPERIENCE OF THE FIVE JAPANESE POLITICIANS MOST LIKELY TO BECOME PREMIER, 1970-1975[1]

Name	*Age*	*No. of Times Elected to Diet*	*Years in Diet*
Takeo Fukuda	67	7	19
Takeo Miki	65	12	32
Masayoshi Ohira	62	7	19
Yasuhiro Nakasone	53	10	24
Kakuei Tanaka	53	10	24

[1] The source for this data is Japan, Shugiin Jimukyoku (House of Representatives, Secretariat), *Shugiin yoran* (House of Representatives Handbook), March 1967 edition (Tokyo: Finance Ministry Printing Office, April 1967).

Another characteristic of faction leaders is long years in the Diet. Table 5 gives the experience of the group of potential premiers. Takeo Miki, for example, has served as long as anyone in the Diet. His experience as a national legislator in the 1930's included making speeches deploring the coming war with the United States. The other men, however, are not the most senior men in their factions, either in terms of age or experience. In fact, Nakasone and Tanaka are still regarded as young men in the party. Fukuda and Ohira are behind the others in parliamentary experience, but each of them had a long career as a bureaucrat in the Finance Ministry before running for a seat in the Diet. On the other hand, all of them have been in the national legislature most of the postwar era. If nothing else, building a faction takes time.

A further characteristic of the faction leaders is that they have all served in the cabinet. Some of the older faction leaders have not regarded this service as important. They have served once, sometimes in a relatively minor post, and have been content to secure seats for their followers thereafter. Among the younger faction leaders, such as the five named above, cabinet experience has been regarded as more important. All of these men have served in the cabinet a number of times, and in most instances the post has been a powerful one. All the faction leaders, then, have had exposure to the difficult business of running a government.

The faction leader's principal characteristic is his ability to maneuver within the party. He is the man called upon to adjudicate disputes. He must supply the political needs of his followers. He must seek out and obtain the political posts they desire. He must see that they receive support during elections. This support includes making sure they have sufficient political funds.

The need for political funds draws the faction leader into close association with business interests. No faction leader could hope to exist for long without fecund sources within the economic community. But while the businessman is important, he cannot dictate to the politician. The businessman is beholden to the bureaucrat through regulation and the implementation of law. The bureaucrat is beholden to the politician as the ultimate source of political power. The conservative power structure, then, is composed of three elements—the businessman, the bureaucrat, and the politician. The politician is clearly the most important.

Thus, the faction leaders head up more than just their own factions. They stand at the forefront of a complex political machine. This machine is responsible and adaptable so long as no undue strains are put upon it suddenly. This requires the faction leader to be predictable in behavior. When the factions first appeared, no one recognized predictability as an important attribute. In fact, when the newsmen found anything to praise about the factions, it was the strongman who had the political muscle to impose his will. No one personified this image of the strongman more than Ichiro Kono. His temper was mercurial. His actions were swift and, to keep the opposition off balance, sometimes slightly irrational.

Kono became a popular hero. He had the support of the media, and he had the skills and a large enough faction to make him a strong contender for the prime minister's chair, but he never made it. He frightened not only the opposition but his allies as well. He was dangerous enough as a faction leader. The businessmen, the bureaucrats, and other politicians were afraid of what would happen if he were to be given the vast powers of the prime minister's office. Other faction leaders have learned from Kono's mistake.

Finally, the faction leaders are remarkably similar to each other. To be sure, each faction leader has his own personality, his own style, and his own interests, all of which create a unique political aura. However, all the faction leaders have walked the same long road and in doing so have been so disciplined that one is much like another, or at least must behave much like another if he is to secure or retain the support of his peers—which success as a premier requires. As a result, replacement of one faction leader by another as prime minister has had little influence on Japan's course in the past, and it is not likely to have much greater significance in the near future.

In view of the impressive stability of the institutional structure of the Japanese political system, the relative balance of the political forces, the very strong presumption in favor of the continuation in power of the conservative party under the same kind of leadership as in the past, and the relative insulation of that leadership from mass public opinion, it must be concluded that over the next five years little change in Japan's security policy can be anticipated—unless there are major changes in the international environment.

CHAPTER IV

The Attitudes of the Business Community

FRANK C. LANGDON

THROUGHOUT the postwar period Japanese business leaders and organizations have participated actively in making and implementing Japanese security policy. As a group they have been among the strongest supporters of the government's fundamental policies of "economism" and "balanced defense," traditionally insisting that the country's top priority go to economic growth and that its security be sought not so much in the build-up of its own forces as in the maintenance of the American guarantee.

Within the last few years, however, new currents have begun to flow. Japan's emergence as the world's third largest economy, the stirring of nationalistic pride, the fear of China's nuclear weapons, and the need in 1970 to reconsider the U.S.-Japan security treaty—these and other new forces have brought to the fore of the business community new voices demanding an acceleration of defense spending.

The arms makers have taken the lead. Closed down by the Occupation, they resumed production on a small scale in the 1950's largely to supply the needs of the American forces in Korea. When the Korean war subsided, they shifted gradually to production for Japan's own Self-Defense Forces. These needs, however, have not been great; and in 1969 Ken Okubo, president of the Japan Weapons Industry Association and chairman of the board of one of its largest components, Mitsubishi Electric, issued a dramatic call for more than quadrupling the defense share of the GNP, raising it from less than 1 percent to 4 percent annually.[1] After all, he argued,

[1] Ken Okubo, "Yagate wa kakubuso o" (Nuclear Weapons Before Long), in *Bessatsu chuo koron: Keiei mondai* (Business Management Problems: A Separate Series of Chuo Koron), Fall 1969, p. 424.

if economically hard-pressed countries like Britain and France can manage 5 percent of GNP and even neutral Switzerland 4 percent, surely Japan, with the second highest GNP in the non-communist world, can afford as much. Apparently thinking almost exclusively in terms of economic capacity, he did not point out that so reasonable an appearing change of the share, in view of the enormity of Japan's economy, would have raised Japan instantly to the rank of third largest military spender in the world and, in relatively short order, the world's third largest military power. He did, however, acknowledge that such a change would raise seriously the nuclear question. He did not call for nuclear weapons immediately, but he did say that he thought the next generation would have to face the question, not because Japan would wish to use such weapons, but "so that it would not be subject to arbitrary treatment by other countries."

From the arms makers, the circle of advocates of accelerated rearmament moves out to the Defense Production Committee of the FEO (the Federation of Economic Organizations, known generally in Japan by its abbreviated title, Keidanren).[2]

The committee plays a valuable part in implementing defense production. The example can be cited here of the Japanese decision under the proposed Fourth Defense Build-up Plan (1972-1976) to produce domestically its own Phantom jets. As with general defense budget questions, this decision

[2] On the committee, see Keidanren, "Boei Seisan Iinkai no katsudo" (Activities of the Defense Production Committee), in *Keidanren jigyo hokoku* (Report on the Work of Keidanren), 1968, pp. 117-20; Yasujiro Okano, "Anzen hosho mondai e no torikumikata ni tsuite" (How to Get Hold of the Security Problem), Keidanren, Boei Seisan Iinkai, June 9, 1969 (mimeo); Keidanren, *Boei Seisan Iinkai junenshi* (A History of the First Ten Years of the Defense Production Committee) (Tokyo: Keidanren, Boei Seisan Iinkai, 1964) privately printed; and a recent report, Keizai Dantai Rengokai, *Showa yonjuyonen Keidanren no katsudo* (Activities of Keidanren, 1965) (Tokyo: Keidanren, December 1969), pp. 37-40.

was reached through negotiations chiefly between LDP politicians and Finance Ministry officials. The discussions involved making rough production estimates and determining the rate at which the planes would become available to the defense forces. When detailed procurement needs for the aircraft began to be outlined by the Defense Agency in October 1969, the Defense Production Committee was called in to help consider, for example, what new factories would be needed, how soon, and what level of future production to expect.

Another example concerns the implementation of naval expansion plans. After several unsuccessful attempts, and amid considerable turmoil, in August 1969 the Diet finally passed two bills dealing with maritime defense and the size of forces. This was a time when business was discussing the possibility of protecting Japan's oil routes. Although the two bills dealt mainly with the Fourth Defense Build-up Plan period, Defense Agency officials met with the Defense Production Committee to consider the industrial implications.[3] It was understood that the maritime defense forces would be given more emphasis and that the number of personnel would be increased. To provide a stronger base for the defense forces in general, defense reserves were to be increased by 70,000 men and new reserves added to the maritime force. Plans were also considered for spending during the final two years of the Third Defense Plan (1967-1971) along with industrial investment for future defense production under the Fourth Plan. Present at the discussion were the president of the Federation of Economic Organizations as well as the heads of Mitsubishi Heavy Industries, Mitsubishi Electric, and the Japan Precision Machinery Company. In attendance from the Defense Agency were the director-general, the administrative vice-minister, the chief of the Secretariat, the Defense Bureau director, and the Equipment Bureau director.

[3] *Nihon keizai shimbun*, August 14, 1969, morning edition.

In view of these intimate relations with the government, the committee's overall views on defense are worth noting. When established in 1953, the committee produced the first comprehensive defense development plan, one more ambitious than anything actually carried out to date.[4] In February 1966, in connection with the government's Third Defense Build-up Plan, the committee proposed specifically the strengthening of the National Defense Council, raising the status of the Defense Agency to a full ministry, and expediting domestic production of Japanese defense material.[5]

It also advocated increasing defense spending annually to 2 percent of GNP, not as much as Okubo has called for, but a doubling of the present share and an amount sufficient to make Japan a major military power by 1975. It has continued to lobby for these objectives ever since.

In February 1967 it proposed an advisory commission to the prime minister designed to popularize the need for more adequate defense efforts.[6] In June 1969 Chairman Yasujiro Okano urged the director-general of the Defense Agency to issue a defense white paper to clarify formally Japan's defense objectives.[7] The government at first hesitated to do so, fearing to stir up criticism, but it did finally act a year later.[8]

Although the Defense Production Committee is cautious in expressing its views publicly, its secretary-general, Tetsuya Senga, who is also a director of the FEO, has offered his personal opinion that Japan should become more independent in the production of all conventional arms.[9]

[4] Keidanren, *Boei Seisan Iinkai junenshi.*

[5] *Asahi shimbun*, February 24, 1966.

[6] *Asahi nenkan* (Asahi Yearbook), 1967, p. 458.

[7] Mainichi Shimbunsha, *Anzen to boei seisan* (Security and Defense Production) (Tokyo: Mainichi Shimbunsha, 1969), p. 71.

[8] Japan Defense Agency. *Nihon no boei* (The Defense of Japan), October 1970.

[9] Tetsuya Senga, *Jishu boeiryoku to Nihon keizai* (Autonomous Defensive Strength and the Japanese Economy), Getsuyokai (The Monday Club) Report No. 445, No. 9 of Security Series, July 14, 1969 (Tokyo: Kokumin Seiji Kenkyukai, 1969), p. 44.

Unlike Okubo, who felt that the government was going to have to take the nuclear decision before long, Senga argued that the need for acquiring nuclear weapons was not so clear. He doubted the seriousness of the threat of China's nuclear capacity and argued that in any event the government should not act unless there were a change in the national consensus. For the foreseeable future, he advocated that Japan dedicate itself to the peaceful development of atomic energy as called for in the Atomic Energy Act. The appropriate goal for Japan, he said, was to become the world's third largest atomic energy producer for peaceful purposes, and, rather than building its own nuclear weapons, it should continue to call on all other states to scrap theirs.

The statements of Okubo and Senga advance three of the arguments most commonly made by those businessmen who advocate increased defense spending: that Japan, with such a booming economy, ought to be able to support as large a share as do other advanced countries; that Japan should become more "self-reliant" in its defense production (Hirotsugu Hirayama, a director of Mitsubishi Heavy Industries, talks of the frustration of the "pride" of Japanese industrialists in not having been able so far themselves to satisfy Japan's defense needs);[10] and that Japan should assume slightly greater responsibility for the defense of its own immediate area.

Two other arguments also are used. One is that often heard in the United States: that, far from retarding the growth rate of the economy, an enlarged military budget would augment it. This would come about, says Chu Kobayashi, for example, as a result of the new technology for civilian industry that could be expected to "spin off" from expanded military production of advanced weaponry.[11] The other is that low budgets mean small orders, small orders mean small production

[10] Hideo Kojima, comp., *Nihon no heiki sangyo* (The Japanese Arms Industry), published by Mainichi Shimbunsha for *Ekonomisuto* (1968), p. 216.

[11] *Ibid.*, p. 215.

runs, and small production runs have higher per unit costs than large ones. In short, the government would get more for its money if it spent more. An enlarged defense budget would also stimulate employment.

It is not easy to estimate the significance of these voices. On the one hand, it is true that defense production in Japan even now accounts for less than 2 percent of the output of Japan's machine industry.[12] By itself, it does not have great weight. On the other hand, it is also true that most of the arms manufacturers are intimately related to the great Mitsubishi, Fuji, Hitachi, Ishikawajima-Harima, Sumitomo, Nissan, Shibaura, and other heavy communications and precision industry combines, which do have clout. And their pleas, while not fully accepted by their combine associates or the wider business community, are beginning to gain ground.

This is particularly true in Nikkeiren, the Japan Federation of Employers Organizations (JFEO). At its general meeting in April 1969 the JFEO declared: "It is natural for us to be essentially self-reliant in the defense of our country. We can say that under present international conditions it is historical necessity that compels us to defend our demands; for pleas for unarmed neutrality [by the Japan Socialist party] and efforts to abrogate the Japanese-American security treaty immediately [by the JSP and the Japan Communist party] are merely ideas that are utterly irresponsible and out of touch with the times. We have now reached the stage where we should enable all the people to push toward an independent defense broadly conceived and expressing the general will."[13]

[12] During the entire Second Defense Build-up Plan for 1962-1966, for example, the share was only 1.76 percent. Japan Defense College, *Boei sangyo ni jittai chosa no kekka* (Results of an Investigation of the Actual Conditions of Defense Production), 1968, p. 8.

[13] "Shin dantai ni tai suru keieisha no taido" (Attitudes of Management toward the New Stage), in *Nikkeiren taimusu*, May 1, 1969, p. 1.

The views of leaders of the Employers' Federation were expressed much more precisely at the top management seminar sponsored by the organization at the Manpower Development Center at Fuji Yoshida in August 1969. In the report by Hiroki Imasato, secretary-general of the federation and president of Japan Precision Machinery Company, the suggestion was made that the security treaty might even have to be revised to accommodate a more equal role for Japan and an increased defense capability. The report stated: "It is undesirable to hold rigidly to the security treaty for a long time. Though an automatic extension of it for the time being [is satisfactory], it is important to continue to seek popular judgment on the problem. It will probably be necessary to make a very thorough reconsideration of it by 1975 in view of the new relationship between Japan and the United States. . . .

"It is necessary to work toward increasing the disposition of the majority of our people to protect their country by themselves and to plan to strengthen effectively our defense potential to the extent that it does not have an adverse effect on economic growth."[14]

Although he did not mention it in his report, Imasato stated during the discussion at the meeting that a defense expenditure equivalent to 1.5 percent of GNP would not retard growth of the economy.[15]

More outspoken is JFEO Director Takeshi Sakurada, adviser of Nisshin Boseki Company. At the October 1968 general meeting of the organization he criticized current defense levels, saying: "I do not want economic development where you have peace and freedom without an assured national defense."[16] A year later he created a sensation by calling for revision of Article 9 of the Japanese constitution, which re-

[14] "Hajime no keiei toppu-semina hiraku" (First High-Level Seminar for Management Begins), *Nikkeiren taimusu,* August 14, 1969, p. 1.

[15] *Nihon keizai shimbun,* August 10, 1969.

[16] *Asahi nenkan,* 1969, p. 465.

nounces the right to wage war, the threat or use of force, and the maintenance of land, sea, and air forces as well as other war potential.

Article 9 was intended to prevent Japanese rearmament, but after the outbreak of the Korean war the government of Japan began interpreting the prohibition against "war potential" as outlawing only offensive forces and operations overseas, not as denying the inherent right of self-defense, a right which, it has argued, justifies the maintenance of a moderate level of defensive, non-nuclear forces for the immediate protection of the home islands. The Self-Defense Forces have been built on this conception. While most Japanese by now have come to accept the need for such forces, many feel morally uneasy with this interpretation of the constitution. Others, particularly among the conservatives, feel it is too restrictive ultimately to serve the national interest. The Liberal Democratic party, consequently, has for years hoped to amend Article 9 expressly to permit military forces for defensive purposes, but it has lacked the necessary two-thirds majority needed to pass such an amendment in the Diet and has therefore tended in recent years to shelve the issue. Businessmen too have said little about it. Sakurada's statement was, therefore, explosive.

Sakurada's statement aimed at reviving the issue to build popular support for an amendment and for defense forces and government defense policy. He said that sensible people expected the government to act with courage to avoid domestic unrest, even though "more serious internal disorder than that faced at present" was likely.[17]

The LDP leadership moved quickly to moderate the reaction. The chief cabinet secretary and the prime minister responded the next day, charging that the statement was "untimely and inappropriate." Opposition party leaders ac-

[17] *Japan Times*, October 17, 1969.

cused Sakurada of trying to step up rearmament, including overseas expeditionary forces and even nuclear arms under the cover of an "independent defense policy."[18] The JSP secretary-general asserted that "big monopoly capitalists" in the Economic Federation, the Employers' Federation, and the Committee for Economic Development had been proclaiming Japan's great economic power since the spring, hoping that Japan would become politically powerful in Asia and even acquire nuclear weapons. He contended that Sakurada was speaking for Japanese "imperialism" and that his statement was just as audacious as Sato's earlier claim that "the main force of Asian stability is Japan."

Sakurada may have stimulated such charges by his call for greater Japanese defense efforts to cope with the shift in the Asian balance of power he foresaw as a result of the American pullback from Vietnam, the British withdrawal from Singapore and Suez, and the Russian advance into South Asia. However, because Sakurada had been deeply involved for years in efforts to build support for the defense forces, his objective may have been relatively modest: to seek to legitimize the self-defense forces rather than to expand its theater of operations.[19]

Quite unperturbed by the hostile reaction to his amendment proposal, Sakurada wrote to one thousand national leaders asking for constitutional revision and a rewriting of Article 9. He argued that the relatively small 250,000-man defense force was generally regarded as illegitimate and that this impaired morale; that the $1.5 billion yearly defense appropriation could not be increased with the constitutional question unresolved; that the purposes of the present defense forces needed clarification; that authorization for them to participate in international peacekeeping was desirable; and that

[18] *Ibid.*

[19] *Mainichi shimbun*, October 27, 1969.

it was irresponsible to be timid about the constitutional revision issue since it ought to be discussed widely and earnestly by everyone.[20]

Sakurada has disseminated his views also through the Defense Discussion Council (Boei Kondankai), organized in September 1965 by a group of business associations and first chaired by the then FEO president, Taizo Ishizaka.[21] Stimulated in part by concern for the successful nuclear weapons tests of the People's Republic of China earlier that year and the year before, the council and its affiliates have devoted themselves to trying to bring about greater public acceptance of Japan's Self-Defense Forces. At the suggestion of Tokusaburo Kosaka, chairman of the board of Shinetsu Chemical Corporation who has since been elected to the Diet, and Masaru Hayakawa, a director of the JFEO, it was decided to organize societies for cooperation with the defense forces (*boeitai kyoryokukai*) throughout the country. A league of such societies was formed at the end of March 1966 in Tokyo under Sakurada.

The council has been active mainly in sponsoring visits to defense force bases and facilities. It stresses the positive contribution that the forces are making. Various schemes have been introduced, such as one enabling private business to send employees to the bases for short training periods, to inculcate a positive attitude toward discipline in a more patriotic atmosphere. The defense forces have even permitted women to come for brief training periods. The cooperating societies have also tried to help former servicemen find employment in industry. Suggestions have even been made for transferring industrial employees to the forces when business is slack and returning them to industry when labor is scarce.[22]

[20] *Japan Times*, October 31, 1969.

[21] *Asahi nenkan*, 1967, p. 458.

[22] "Jishu-boei no senryaku koso" (Strategic Concepts of Autonomous Defense), *Toyo keizai*, September 6, 1969, translated in U.S.,

Moreover, the interchange of top personnel between the Defense Agency and business has also been considered, although it is feared that such a move would be attacked by critics as flagrant evidence of the existence of a dangerous "military-industrial complex."

The council also has assisted in promoting contacts between American and Japanese businessmen concerned with defense production. A similar defense council in America with headquarters in Los Angeles sent two delegations to Japan in 1968 to discuss defense production plans. They were said to have spent more time with the Japanese Defense Discussion Council than with government officials or agencies.[23]

The American delegations of industrialists were interested in sounding out Japanese business views on Japan's coming defense requirements, especially for the Fourth Defense Plan on which some major decisions were being made. Visiting Defense Department officials from the United States also contacted the Japanese council.

In 1969, upon the invitation of the American council and the Defense Department, the Japanese Defense Discussion Council sponsored a mission made up of heads of leading Japanese defense-related firms to go to the United States from October 29 to November 11 to inspect defense installations and the Houston Space Center, as well as to talk with managers of Lockheed, General Electric, Boeing, and McDonnell-Douglas. An important objective of the mission was to explore the possibility of cooperation between the companies of the two countries in both defense production and space development. Even where the Japanese hoped to produce

---

Department of State, *Summaries of Selected Japanese Magazines* (Tokyo, American Embassy), October 6-20, 1969, p. 21.

[23] *Sankei shimbun*, September 30, 1969, translated in U.S., Department of State, *Daily Summary of Japanese Press* (Tokyo, American Embassy), October 1, 1969, pp. 18-19. See also *Asahi shimbun*, October 10 and 29, 1969, morning editions.

their own military aircraft, they would be dependent for much of the necessary technical information upon the willingness of American business to cooperate. The mission was to visit Washington and talk to defense officials and businessmen. An important objective was to learn as much as possible of American views on the Southeast Asian military situation in order to estimate how Japanese defense plans might be affected in the 1970's.

But the advocates of accelerated rearmament have not yet captured the citadels of the business community. The Federation of Economic Organizations, which is the single most important business organization in Japan, sticks closer to the government than to its own Defense Production Committee on this issue. This is significant because the FEO participates in official policy-making via ministerial advisory committees and through petitions and consultation with government and party officials. It represents the consensus of business opinions favoring increased defense capability, continued adherence to the security treaty, and greater responsibility for Southeast Asian stability. Officers of the organization are bankers as well as industrialists, so that its position is perhaps the maximum that can be attained in the way of agreement across business divisions.

Leaders of the federation have generally avoided any reference at major gatherings to defense problems, so it was quite unusual on May 23, 1969, when President Kogoro Uemura went even so far as to say what the premier had been saying:

"The end of hostilities in Vietnam continues to be likely, but imminent British withdrawal from Southeast Asia presents big problems for peace and stability there. It is Japan's duty to continue the security system [the security treaty and U.S. bases], to increase gradually our independent defense strength, and under a collective security system to contribute to the maintenance of peace in the Far East. This is an international duty and an important issue concerning Japan's own future prosperity."[24]

Even this cautious assertion is too strong for another of Japan's powerful business associations. The leaders of the Japan Committee for Economic Development (Keizai Doyukai) champion the view that is perhaps the most widely held in the Japanese business community: that Japan should pursue a foreign policy of peaceful economic internationalism, as non-military as possible. Rather than an ordinary business association, the committee is actually a club of prominent business heads who take a relatively liberal attitude on labor and political problems. The committee has produced a number of moderate spokesmen who have been particularly opposed to reliance on military power and who feel that economic cooperation is the key to regional influence. Its chief officer is Kazutaka Kikawada, chairman of the Tokyo Electric Power Company, whose views are representative of the group's thinking. He has said: "Although some self-defense power is necessary, it should be exercised within the scope of the peace constitution, which should not be amended. We should not believe in Japan replacing the United States, which is about to withdraw from Asia. Cooperation in stabilizing the livelihood of the various Asian peoples is the shortest way to peace."[25]

Kikawada's goal is to work toward world peace without excessive military preparations and to continue present types of collective security only so long as necessary. One influential committee member, Fuji Bank President Yoshizane Iwasa, puts it: "Japan's security policy must be broad enough to include diplomatic and economic means in addition to some means of defense."[26]

[24] *Asahi shimbun*, May 24, 1969.
[25] *Mainichi shimbun*, October 27, 1969.
[26] *Ibid.*

Another moderate spokesman is Tokusaburo Kosaka, who was elected for the first time to the House of Representatives in the December 1969 general election to become one of less than a handful of top business leaders to enter the Diet since the war. His brother, Zentaro, is head of the LDP's foreign affairs committee and served as foreign minister under Prime Minister Ikeda. At the final United States-Japan Assembly held at Shimoda in September 1969, where scholars, businessmen, and politicians from both countries met to discuss diplomatic relations between the two countries, Tokusaburo Kosaka stated that the idea of an "independent defense" then being advanced in conservative political circles should not be thought of as being the same as defense solely by Japan's own efforts. He felt that leaders of the Employers' Federation tended to equate the two ideas, and that such thinking led logically to the acquisition of nuclear weapons. Seeking apparently to appropriate the nationalist appeals of the hawks while adhering to traditional policy lines, he argued that the best "independent defense" for the present should mean the capability to defend Japan with conventional weapons, coupled with a defense arrangement with the United States.[27]

Business moderates emphasize economic over military methods but are not averse to increased defense capability in conventional weapons, within limits not disturbing to the economy. Kikawada, for example, recognizes the need "to increase defense power enough so that Japan can cope with a local war and protect commerce in the surrounding areas." He also attaches importance to achieving "a balance with the national economies of other countries."[28] Another businessman active on the committee, Shigeo Nagano, chairman of Shin Nihon Steel and head of the Japan Chamber of Commerce and Industry, has declared: "There is a limit to a foreign country's carrying out attacks with conventional

[27] *Mainichi shimbun,* October 27, 1969, morning edition.
[28] *Ibid.*

weapons. Japan has only to maintain a defensive strength sufficient to cope with this and, even in this case, it is desirable [in increasing defense expenditures] to take enough time so that there will be the least possible impact upon the standard of living of our people."[29] Heigo Fujii, vice president of the Committee for Economic Development and vice president of Shin Nihon Steel, is also cautious and recognizes the limits imposed by Japanese public opinion on defense increases when he states: "Policies for national defense must be conducted strictly through popular consensus."

Even in moderate or internationalist thinking, nationalist feeling is expressed in support for Japan's playing a larger role in Asia and more equality and independence vis-à-vis the United States. Dovish businessmen stress, however, that Japan's contribution will be economic, not military. As Kikawada puts it: ". . . In order to cooperate in dollar defense and also to advance economic cooperation in Southeast Asia, we should be equal partners [with the United States]. . . . I think the international division of labor from now on should be that the United States handle military affairs and Japan economic cooperation."[30]

On balance, it would seem safe to conclude that for the foreseeable future, at any rate for the next five years, barring any overriding changes in the distribution of world power, the Japanese business community is prepared to support a modest stepping up of the defense effort, including an increase in GNP share to somewhere between 1 and 2 percent a year, a greater emphasis on self-reliance in the production of Japan's arms needs, and a gradual assumption of greater responsibility for the military defense of Japan's own immediate area, looking forward perhaps in the late 1970's, but certainly not before that, to a phasing out of the security arrangements with the United States. But it is not pressing hard

[29] *Ibid.*

[30] Mainichi Shimbunsha, *Anzen to boei seisan*, p. 425.

for even these modest changes. It is clearly not calling for any massive rearmament. Nor is it prepared to grasp the nuclear nettle, though it is reluctant to foreclose its right to do so by ratifying the Nuclear Non-Proliferation Treaty.

In the past few years there has been an upsurge of interest in the subject of Japan's acquiring nuclear weapons. Businessmen are now well aware that Japan has the economic capacity to enter this field if it chooses. Herman Kahn notes that Japan's nuclear power reactors for the generation of electricity will have as a by-product enough plutonium to produce five hundred to a thousand small nuclear weapons a year in the late 1970's.[31] Minoru Genda, a Diet member and former chief of the air defense staff, recently said that Japan could easily produce a hundred warheads of the Hiroshima type in less than ten years at a cost of only $200 million. But even among those businessmen like Okubo who anticipate that Japan will take that choice eventually, no responsible leader is calling for them now.

They are constrained partly by Japan's international dependence in this field. The Atomic Power Company's Tokai Mura station is designed to produce 166,000 kilowatts of electricity and the Suruga reactor is to produce 322,000 kilowatts. Both plants were in operation by the end of 1969. The Tokyo Electric Power Company's Fukushima Plant Number One and the Kansai (Osaka-Kobe-Kyoto region) Electric Power Company's first plant were to commence operation in 1970. In 1969 Japan had only 170 megawatts of nuclear power compared to West Germany with 900 and Italy with 620. By 1975, however, Japan should have 6,000 to 7,000 megawatts, which will be a large program.

But Japan depends completely upon foreign sources of fuel, technology, and, to some extent, even investment capital

[31] Herman Kahn, Anthony J. Wiener, et al., *The Year 2000, a Framework for Speculation on the Next Thirty-Three Years* (New York: Macmillan Co., 1967), pp. 244-45.

for its nuclear development program. There is no uranium-bearing ore in Japan of any importance, so six Japanese mining companies and the Tokyo and Kansai Electric Power companies in collaboration with an American company are prospecting for the ore in Canada, Australia, Africa, and South America. Nor does Japan produce its own fuel. The Research Institute of Physics and Chemistry under the Atomic Energy Commission is engaged in fuel processing by the gaseous diffusion method, while the Atomic Fuel and Power Authority is investigating the centrifugal separation technique to produce enriched uranium. However, Japanese reactors are still dependent upon supplies of enriched fuel from the United States.

Capital for power plant construction has come from the United States Export-Import Bank. Japanese bankers and industrialists as well as the government have thus far put only limited resources into producing their own equipment. Japanese firms planning to manufacture atomic equipment are working separately with different American and British firms and thus are competing with each other for the prospective Japanese market. Moreover, unlike other advanced countries, Japan has divided its limited research funds in an effort to develop two different types of reactors. Consequently, there has been a splintering of activity in Japan's own reactor and fuel processing development. Government aid and leadership have been minimal, and business has been unable to bring order into the situation.

Businessmen are sensitive also to the view of some government advisers that possession of nuclear weapons on a lesser scale than that of the United States and the Soviet Union might be worse than having none at all. A small arsenal would merely stimulate nuclear competition with other Asian states such as China and India and arouse fear among the rest. This would jeopardize Japan's new peaceful image in Asia and the Pacific and threaten its sources of supply and its export markets, built up at such tremendous cost and effort over the postwar era. Access to essential raw materials is an important consideration in maintaining this peaceful image. As Yoshiro Inayama, president of Shin Nihon Steel, comments: "We now buy a large quantity of iron ore from Australia cheaply. For a long time after the war Australia felt hostile toward Japan and would not sell its ore to us. But it is unthinkable that Japan, as a peaceful state, might again become a threat, so the feeling toward us has changed. It has become possible to buy. Peace is, therefore, the major precondition without which we cannot carry on economic development."[32] Domestic tranquillity also is essential to business, and the public's antipathy to nuclear weapons is such that most businessmen prefer not to provoke it.

Thus, by relying on the American deterrent and without nuclear weapons of its own, it is felt, Japan can devote itself to business. Of course, this thinking is behind the dominant postwar Japanese foreign policy of peaceful commercial expansion protected by rearmament to insure American defense of Japan—a policy inherited from the Yoshida Cabinet twenty years ago.

On the other hand, the business community is unanimously in favor of the peaceful development of nuclear energy. Necessary primarily to meet Japan's enormous anticipated power requirements in the near future, it may also serve to insure Japan's continued economic progress because of the technical spin-off associated with research and development in this field.

One controversial issue regarding nuclear weapons concerns Japan's participation in the Nuclear Non-Proliferation Treaty. This is one of the surprising cases where Japan has opposed the United States on a non-commercial question and also, apparently, has gone counter to its basic policy of opposition to acquisition of nuclear weapons. Even more sur-

[32] Mainichi Shimbunsha, *Anzen to boei seisan*, p. 141.

prising has been the unanimous agreement of the opposition political parties in rejecting adherence when they have always claimed to be the most dedicated in their opposition to nuclear arms. The clue to this behavior seems to be the arousing of nationalist feeling.

The treaty is felt by the Japanese to be inherently unequal in its impact, as there is no binding limitation on signatories that already have nuclear weapons to disarm or reduce their nuclear arms. It is especially onerous for countries like Japan, which can most readily undertake nuclear armament, to pledge to forego this option for the future. Government and business leaders have also been troubled by other possible effects involving peaceful nuclear development. Their objections have related principally to the unequal nature of the inspection system: that is, international inspection not of the possessing countries, but of those like Japan who are engaged in developing their own fuel processes and reactors. Any opportunity to advance in nuclear development might be lost to Japan, it is argued, by the access of international inspectors to its nuclear research and production facilities. Japan might be further disadvantaged by International Atomic Energy Agency inspection in comparison with Germany, which will be inspected by EURATOM, an agency to which it already belongs and which contains more friendly participants. A still further disadvantage is the prevention of nuclear explosions even for peaceful purposes. While Japan can arrange to pay for such explosions by the advanced nuclear nations not bound by the treaty, it may be difficult to benefit fully from international nuclear development, especially when some military aspects are forbidden by American law to be revealed, even though the United States government has offered to share the peaceful spin-off from military technology.

Japan at present has no such secrets to lose. On the contrary, it needs to import all the technology it can get. As Ryukichi Imai, consultant to the Japanese Foreign Ministry and associate of the Japan Atomic Power Company, has pointed out, Article 5 of the treaty promises the sharing of applied nuclear information and Japan needs to press for the technological information behind it. Dependence in industrial technology can be a very restrictive relationship, however, and runs against the Japanese desire for national independence.

Perhaps a greater objection is the fear of industrial dependence and control. There is the fear that the military and nuclear dominance of the two superpowers is being perpetuated in the areas of nuclear development and nuclear power generation. Until Japanese government and business leaders really get behind Japan's nuclear development and create their own fuel processing system and their own reactors and locate their own source of uranium free of great power control—let alone acquire the accompanying technology—there is little hope of commercial nuclear independence.[33]

It may be the very realization of the extent to which Japan is dependent upon the United States that makes Japanese business want more genuine independence, in keeping with its industrial output, but hesitant to bite the hand that feeds it. This desire for independence plus the fact that it is politically undesirable to appear to oppose nuclear disarmament has led to a certain ambivalence toward the treaty on the part of Japanese business. Generally, business reaction has been cool. As the Imasato report presented at the Employers' Federation seminar in August 1969 suggested: "On the question of the non-proliferation treaty it is necessary to proceed with particular caution. With respect to the possibilities of developing and manufacturing nuclear fuel it is important to keep enlarging Japan's voice in the matter internationally."[34]

[33] See Japan, Cabinet Secretariat, Naikaku Chosa Shitsu (Cabinet Research Office), "Nihon no kaku seisaku to gaiko" (Japan's Nuclear Policy and Foreign Relations), *Chosa geppo* (Monthly Report of Research), No. 150, June 1968, p. 7.

[34] *Nikkeiren taimusu*, August 14, 1969, p. 1.

After West Germany and Italy had signed the treaty and it was about to come into force, Japan signed it on February 3, 1970. The government, however, like Germany, said that ratification would not take place until inequalities of international inspection were eliminated and peaceful nuclear development was not hampered by inequality between the nuclear weapons states and non-weapons states. These reservations were very substantial indeed and had business as well as government support.

One characteristic of all but the most hawkish of businessmen is their strategic dovishness. Even Senga calls for the use of Japan's arms only for the defense of its own home islands, including Okinawa, and the immediately adjacent waters.[35] He does argue that Korea merits Japan's security attention but would have Japan limit its military action to deterrence only, in no circumstance sending its own combat forces as the United States did under UN auspices in 1950.[36] Beyond that, there is almost no business support for Japanese military protection of its overseas interests.

During the nationwide defense debate of 1969-1970, the question was occasionally raised whether Japan should plan to try to defend the sea routes of its indispensable oil supply. Oil occupies 70 percent of Japan's energy sources at present. As much as 90 percent of Japan's crude oil imports comes from the Middle East via the Indian Ocean and the Macassar Straits; it imports practically no refined petroleum products. By remaining strictly neutral, Japan did manage to keep oil imports coming from Muslim countries during the 1967 Arab-Israeli conflict, but the closing of the Suez Canal and the subsequent increase in tanker rates had serious repercussions for the Japanese economy. Business is concerned now over the possibility of further hostilities in the Middle East, the operations of the Soviet navy in the Indian Ocean, the

[35] Senga, *Jishu boeiryoku to Nihon keizai*, p. 44.
[36] *Ibid.*, pp. 41-42.

withdrawal of U.S. forces from the area, and the spread of communist strength in Southeast Asia. This concern has led to a desire for a greater emphasis upon maritime defense in the projected Fourth Defense Build-up Plan for 1972-1976. Some businessmen, recalling that it was the cutting off of Japan's oil supplies in Southeast Asia that precipitated the Pacific war, have even wondered whether Japan should develop a capacity to escort its tanker fleets at least as far as the Straits of Malacca. One estimate of what this might require is supplied by Tomoharu Nishimura, former chief of the Maritime Self-Defense Force staff and advisor to Kawasaki Heavy Industries Company.[37] Nishimura says that without American cooperation it would take a navy three times the size of the present Maritime Self-Defense Forces. Incidentally, he suggested that the route might have to be shifted to the Lombok or Macassar Straits since ships over 300,000 tons cannot navigate the Malacca Straits. In any event, there is little business support for taking on such an extended defense obligation. The view has come to prevail that it is beyond Japan's capacity and, furthermore, would only antagonize the nations of the region. Most agree with Kikawada that the best way for Japan to protect its interests there is "by economic cooperation which contributes to the improved welfare of the inhabitants of the region rather than by military force."[38]

In practice, this means to insure access to raw materials by extending economic aid to the governments of the region and making credits available to companies engaged in extraction. This applies to the Arabian Oil Company and the North Sumatra Oil Company, for example, which are developing overseas crude oil sources. In addition, joint ventures have been established between Japanese and foreign businessmen, not only in the developing countries, but in the United States,

[37] *Summaries of Selected Japanese Magazines*, October 6-20, 1969, p. 20.

[38] Mainichi Shimbunsha, *Anzen to boei seisan*, p. 144.

Canada, and Australia as well, in order to guarantee future supplies against competition from other advanced countries.

This same "little Japanism" is reflected in business attitudes toward arms sales abroad. At present Japan produces arms almost solely for its own use. The Ministry of International Trade and Industry (MITI) regulates arms exports in accord with its so-called "three principles": no sales to communist countries, countries embargoed by the UN, or belligerents. While this policy has virtually cut off arms sales abroad, it does not follow that the Japanese arms industry is constitutionally opposed to exporting its products. As already noted, it got its start in the 1950's by producing for the U.S. forces in Korea. In the late 1950's Uemura, who is now president of the FEO, sought to build on this by offering what came to be known as the "Uemura Plan" for selling arms to Southeast Asian countries by a sort of barter arrangement. Under the plan the United States was to be asked to send new weapons to Japan. These would be paid for in yen, which the United States would then offer as aid to Southeast Asian countries to buy arms for themselves in Japan. The proposal was rejected by the National Defense Agency, which feared to disturb the American military aid grants that Japan was then receiving.

The arms makers continued to look longingly at overseas sales. They are inclined to agree with Osamu Kaibara, executive secretary of the Defense Council, the government's top military policy body, that Japan is the "natural arsenal of Asia."[39] And they agree with Hirayama that it is economically wasteful for Japanese industry not to be utilized more fully. Fear of antagonizing public opinion at home and abroad has kept them from clamoring too loudly, and Japan's trade balance has been too favorable for their economic arguments to gain much attention. Thus, despite the enthusiasm of the arms industry, the general Japanese business community is unlikely

[39] Kojima, *Nihon no heiki sangyo*, p. 218.

to favor any large export of arms, at least through the early 1970's. But on a small scale, exports are likely to begin. In the summer of 1971 the first request for an export license was being tendered MITI by Kawasaki Heavy Industries, who wished to respond to a Swedish navy order for large helicopters equipped for rescue operations. Other conversations were in progress between the Shin Meiwa Industries and the U.S. navy, and between the Mitsubishi Heavy Industries and the Swedish air force.[40]

The central assumption in all these minimal military conceptions is that the United States can be relied on to secure the larger international environment that the Japanese economy requires. In fact, no one in Japan has been a stronger supporter of close cooperation with the United States than the businessman. He has seen it as essential, particularly in the realm of technology, trade, and finance, and only slightly less so in the field of military defense. He supported strongly the security treaties of 1951 and 1960. In the defense debate of 1969-1970 he generally backed the government's policy of automatic extension; and there is little evidence that even those few hawks who call for an acceleration in the arms build-up can conceive of doing without the security treaty at the present time. Probably most of the business community would agree with the JFEO when it called immediate abrogationists "irresponsible." But probably most would agree also with the JFEO's secretary-general, Imasato, when he said: "It will probably be necessary to make a very thorough reconsideration of the treaty by 1975."

[40] *Asahi shimbun*, August 6, 1971.

CHAPTER V

The Confrontation with Realpolitik

DONALD C. HELLMANN

TO AN EXTRAORDINARY EXTENT during the past two decades, Japan's international role has been reactive, defined almost entirely by the outside environment.[1] Moreover, continuity within the domestic political system should assure perpetuation of this passivity in the immediate future. It is to the external environment, then, that one should look for stimuli. And here for the 1970's one finds stimuli of the most provocative kind: confrontation for the first time in the post-war era with the realities of world politics.

One key to the derivative nature of Japan's past international actions lies in a decision-making process that has prevented bold leadership. This has been in large part due to cleavages in and among the political parties and the lack of consensus on certain basic foreign policy goals. The vehement and unqualified opposition of the Japanese Left to any move toward a more activist position beyond "disarmed neutralism" has discouraged government initiatives by increasing the domestic political risks that would be involved. In addition, continuing conflict in and among the factions of the conservative Liberal-Democratic party, which has held power since independence, has impeded innovations by the prime ministers and deterred decisive moves by projecting intraparty politics deeply into all issues of foreign affairs. The result has been a kind of policy immobilism. Barring an unexpected re-

[1] For elaboration of this point see Donald C. Hellmann, *Japanese Domestic Politics and Foreign Policy* (Berkeley: University of California Press, 1969), especially Chapters 1 and 9. For a more extended discussion of other themes in this chapter, see Hellmann, *Japan and East Asia: The New International Order* (New York: Frederick H. Praeger, forthcoming).

versal of past trends altering the relative strength and modes of operation of the parties or the sudden emergence of a nationalistic consensus, this style of policy formulation will continue to restrain leadership no matter what the personalities or issues of the moment. Further, it will proscribe a Gaullist-style move to develop a more autonomous political and military rule in the world and will insure that any basic shift in policy direction is closely tied to changes in the international milieu, not independently initiated by policy-makers in Tokyo.

Both the extraordinary success of Japanese "peace and prosperity" foreign policy and the pacifistic dimension of the foreign policy debate work to prolong the international passivity of Japan. In 1972, Japan possessed a level of economic well-being and international standing beyond all past expectations, while enjoying peace in a world of turbulence. This has been accomplished by abdicating all concern for the "high politics" of power and prestige, by abandoning the military dimension of diplomacy, by "separating" politics from economics—in a word, by pursuing foreign affairs more like a trading company than a nation state. No matter how inappropriate a foreign policy with such limited parameters may become in terms of external realities, the momentum of past success will inhibit new initiatives, especially in the military field. *Pari passu*, the aims and skills of statesmen and those of international entrepreneurs are far from identical. Pacifism, legally sanctified by the constitution, enjoying wide popular support, and reified as an "inalienable right" by the Japanese Left, still has sufficient political weight to deter any sudden swing toward rearmament within the very immediate future, no matter how pressing the international exigencies. This in addition increases the inertia of Japanese foreign policy and thereby raises the potential for a widened gap between national security capabilities and national needs brought on by altered external conditions. In these ways, the political mood and goals of Japanese foreign policy impede even the acceptance of the imperatives of *realpolitik* and thereby further insure that Japan's basic international role will remain essentially an echo of external developments. Except in the improbable event that international conditions will permit Japan's continued full withdrawal from power politics, this echo will increasingly include the sounds of armaments. Both the timing and intensity of these military reverberations ultimately depend on the now equivocal nature of American policy toward Asia.

The alliance with America has served as an international incubator, insulating Japan from the war and upheaval that have convulsed the Asian region and fostering imbalanced national development by stultifying capacities to cope independently with conflict-dominated regional conditions. Now the protective screen is being lifted, for whatever the ultimate meaning of the Nixon Doctrine, or however favorably disengagement from Vietnam is achieved, uncertainties regarding the ultimate commitment of the United States to security throughout the East Asian region have been raised. Despite preoccupation in both nations with day-to-day diplomacy, fundamental changes in the dynamics of international politics in East Asia will unfold during the next five years. American and Japanese formal assurances notwithstanding, Japan's security position has basically altered and for the first time since 1945 must more and more directly confront the broader issues of war and peace in East Asia. For the incubator infant, the difficult process of adjustment to the real world ultimately depends on the stimuli of his environment, and these can be controlled and softened through parental care. For Japan, the process of adjustment to the harsh realities of the international world should prove even more difficult, for the vicissitudes of the external environment are not only severe but demonstrably beyond the control of the United States. Whether Japan will be stimulated into adding a military dimension to its diplomacy depends on the specific nature of the problems that emerge within the two international systems of which it is a member: the one global, the other regional.

Three essential aims have underlain Japan's participation in the global international system: (1) development of economic relations with all nations while gaining acceptance within the club of advanced Western industrial powers; (2) achieving a respected status in the world international order, primarily through activities in the United Nations and other international organizations; and (3) cultivating the alliance with the United States to provide for Japan's security and to fuel national economic development. Japan has succeeded remarkably in fulfilling what are truly modest goals during a period of revolutionary change in international politics. Now, however, the very enormity of Japanese economic success has brought new problems of international competition with the United States at a time when bilateral relations are in a state of transition and world politics is moving toward a more complex multipolar system.

The American alliance continues to be the critical link for Japan to the global system and the most important outside determinant of Japanese foreign policy. A prefatory summary of how this alliance has evolved is essential to any evaluation of the impact of international politics on Japan.

Building on the opportunities provided during the six-year Occupation, bilateral political and cultural contacts have remained uniquely close. At this level, the relationship has had a paternalistic cast, with the United States serving as a kind of political-cultural stepfather. One result has been the development of elaborate and highly effective channels of formal communication that have had notable success in anticipating and diffusing potential conflicts in the past and can serve as a powerful instrument for deterring any future autonomous security moves at odds with the desires of the United States.

The successful and self-conscious American effort to broaden contacts with the Japanese people, in keeping with the democratizing emphasis of the Occupation, has ambiguous implications for future relations with the United States. Although the reservoir of good will that has been nurtured will impede any sudden rupture in American ties, at the same time the pervasive presence of the United States has led to a kind of bilateral myopia in the foreign policy debate. Today, at the end of this period of prolonged intimacy, conservatives and leftists alike commonly measure international autonomy and nationalism in terms of movement away from the American axis. The relative importance of the links forged as a result of the special political circumstances of the last twenty-five years becomes clearer against the backdrop of economic and security relations, where questions of self-interest and competition are more clearly involved.

Japan is so heavily dependent on trade with the United States that this alone places an important limit on any sharp change in relations between the two countries. In 1970 the United States took approximately 31 percent of Japanese exports, but this was only 14.4 percent of total American imports. Similarly, although the Japanese relied on the United States for 29.4 percent of their imports, this constituted only 11 percent of American exports.[2] Although the volume of bilateral trade, $10.5 billion, is impressive and Japan is our largest overseas trading partner, the asymmetry makes it inappropriate to describe this relationship in terms of interdependence. A decision by the United States regarding trade may have a sizeable impact on Japan without seriously disrupting the American economy. This provides the United States with substantial leverage for pressure on political as

[2] International Monetary Fund, *Direction of Trade*, March 1969; and Japan Customs Association, *Boeki nenkan* (Trade Yearbook), 1970.

well as economic matters, but it also raises the possibility of serious bilateral conflict generated by successful lobbying of narrow interest groups in the United States, such as the textile industry. Japanese business is further tied to the United States through a large number of technical and licensing agreements, particularly in security-related industries. Thus, despite Japan's sharply increased capacity for international action as a result of economic growth, dependence on the United States market will constrain any radical *realpolitik* maneuvers.

At the same time, the potential for conflict will expand as Japan emerges as a major competitor of the United States in both international and domestic markets. This has already been manifested in the controversy regarding the import of Japanese textiles, which has befouled the atmosphere of bilateral relations since 1969, and in the bitterness surrounding the huge bilateral balance of payments deficit (between one and two billion dollars annually) that the United States has incurred since 1968. Both of these issues directly contributed to the drastic measures initiated in the summer of 1970 by President Nixon to rectify America's deteriorating international trade position and to strengthen the dollar by forcing global monetary changes—actions which seriously eroded overall relations between the two countries. The inseparability of politics and economics will be underscored by Japan's very success as an international commercial competitor.

The foundation of the American-Japanese security alliance, as well as the attendant military arrangements, grew out of the vastly expanded commitments of the United States in East Asia resulting from the Cold War. Past United States security policy toward Japan has been determined by the general policies designed to contain communism in East Asia, with bilateral considerations accorded secondary importance. This was clearly manifested in the 1951 peace treaty in which the central objective came to be not settlement of issues related to the Pacific war but securing an alliance with Japan to check communist expansion in Asia.[3] The accompanying security treaty provided that the American troops stationed throughout Japan could be used in any way that Washington felt they would "contribute to the maintenance of international peace and security in the Far East." Revision of this treaty in 1960 occurred prior to the Sino-Soviet split and was again predicated on Cold War assumptions that the communist monolith posed the security threat in Asia, which the United States bore the full responsibility of meeting. Although the revised treaty, which continues in force with option of annual review for both parties, emphasized reciprocity and partnership, Japan remained effectively an American military protectorate and a crucial link in United States defense strategy in Asia.

During the past decade this one-sided military partnership has been supplemented by American efforts to prod the Japanese into a leadership role in Asian economic and political affairs. This policy assumed that the United States would continue to remain deeply engaged in the region both on nuclear and conventional levels, that there was and would continue to be a basic identity of Japanese and American security interests, and that the economic and political dimensions of international politics could be separated from security matters. All of these assumptions have been subjected to serious challenge as a result of revisions in American policy toward Asia in response to the bitter domestic political divisions and the spiraling material and international costs growing out of the war in Vietnam.

On the one hand, the enormous military and political investments of the United States in Vietnam have created the kind of implausibly exorbitant commitment to Asian security implied in our policy toward Japan. This has prolonged the Japanese withdrawal from all concerns of *realpolitik* and has

[3] Frederick S. Dunn, *Peacemaking and the Settlement with Japan* (Princeton: Princeton University Press, 1963), pp. 45-52.

allowed the measured but steady expansion of their own conventional forces to occur free from any tangible security threats. On the other hand, the stalemate in the war has demonstrated that the United States, despite its vast diplomatic, economic, and military resources, lacks the capacity effectively to control international conflict in the region. This, together with the increasingly isolationist mood that has emerged in the wake of the divisive internal political debate over the war, makes highly improbable any American military overtures in East Asia in the near future. Thus, Vietnam at once insured the short-term success of U.S. policy toward Japan and demonstrated the unfeasible nature of America's general Asian policy involving a policeman-like obligation toward conflicts in the region. Because Japan has been unequivocally and consistently considered as integral to regional security and because the Japanese are now so fully involved in East Asia on all levels but the military, this alteration in the intentions of U.S. security policy has special significance for the American alliance and Japanese rearmament. Pressures for regionally centered security arrangements have been raised both by the United States and the countries in the region. Any arrangements of this sort can have long-term meaning only with the active participation of Japan. Consequently, despite Tokyo's clear awareness of the maelstrom of East Asian politics and the strong internal political pressures opposing military involvement abroad, external forces, created primarily by the change in U.S. military policy toward East Asia, are pulling Japan into the international vortex that it has so scrupulously avoided.

Gradual American disengagement from Asia is part of a general global trend toward multipolarity, encompassing limitations on Soviet commitments in Asia and the emergence of a defiantly autonomous and nuclear foreign policy by the People's Republic of China. The Sino-Soviet split has not only induced a kind of Peking-centered regionalism within the Asian Communist world[4] but has also led the Chinese and Russians to pursue distinct and conflicting policies toward non-communist nations in East Asia. Moreover, the Chinese, by virtue of their size, proximity, national ambitions, and place at the "vanguard of revolutionary forces," are viewed by the nations of the region as the major threat to their external security and internal stability. China has become such a rival of the Soviet Union that the major incentive for increased Soviet involvement in the political and military quicksands of the underdeveloped countries of East Asia is to check Chinese expansion. From the perspective of Moscow as well as Washington, the era of unqualified bipolar globalism has passed; and Soviet short-term tactics and long-term goals regarding East Asia are today even more obscure and uncertain than those of the United States. What has developed on the security level is a kind of de facto regionalism with diminished roles for the superpowers and greater importance for local (i.e., regional) matters of defense. This has enormous direct and indirect importance for Japan. Basically, the rift between China and Russia has rendered obsolete the Cold War strategic premises of the Japanese-American alliance, despite the fact that this has not yet been reflected in the formal arrangements between the two countries. Although China's rapid progress toward becoming a major nuclear power cannot be translated into a global capacity to act in the immediate future, the strategic importance of such weapons is immediately manifest on the regional level, where Peking can act against equal or lesser powers. The threat has been felt most strongly by Japan and India, where the pressures for nuclear parity have steadily grown and ratification of the Nuclear Non-Proliferation Treaty has been postponed. From the

[4] China holds a position of comparable influence to the Soviet Union in North Vietnam and North Korea; and all opposition communist movements in East Asia, except the Japanese, are dominated by groups allied with Peking.

viewpoint of Tokyo, the major security threat has shifted from Moscow to Peking,[5] from the still militarily bipolar global system to the East Asian regional subsystem, at a time when Japanese security relations with the United States are being seriously questioned for the first time since 1945.

The very success of Japan is achieving a high level of economic growth and a corresponding status among the developed nations of the world is directly connected with a trend toward multipolarity. Much has been written about Japan's "economic miracle," but only recently has attention been drawn to the significance of the increased capacities for international action that this provides. Japan, now economically the third largest nation in the world, is increasingly seen as the only prospect during the next decade for a "new pole of military power, independent of American preponderance and capable of affecting the regional balance of power."[6] Some understanding of this mounting concern and the relative enormity of Japanese economic development during recent years is provided by noting how fully the positions of the Chinese and Japanese economies have been reversed. In 1957 the GNP of the People's Republic of China was more than one and one-half times that of Japan, but by 1969 Japan's economy had grown to be approximately three times as great as China's (Table 1). Similarly, dramatic changes have occurred regarding the global economic powers, and these will become even more pronounced in the immediate future if past patterns of growth continue. As seen in Table 1, if the economies of the major world powers expand at their expected rates, by 1975 the absolute size of the Japanese economy will have grown to two-thirds of that of the Soviet Union, a figure roughly equal to that of the Common Market today. Most importantly, the

[5] For a different estimate of Japan's perception of China, see Kenneth Young's treatment in Chapter VI, pp. 184-86.

[6] See, for example, Robert E. Osgood, "Reappraisal of American Policy," in Robert E. Osgood, et al., *America and the World* (Baltimore: The Johns Hopkins Press, 1970), p. 23.

gap between Japan and West Germany will have become so large ($140 billion), that Japan will more appropriately be classified with the economic superpowers. Thus a world *à trois*, heretofore linked almost exclusively with the potential of a united Europe, will come into being in an economic sense, marking a fundamental change in the structure of the post-1945 world order.

When and how this shift from economic bipolarity to tri-

TABLE 1

GROSS NATIONAL PRODUCTS OF LEADING GLOBAL AND EAST ASIAN NATIONS
(In Billions of U.S. Dollars)

	1957	*1969*	*1975 (Projected in 1969 Dollars)*[1]
United States	443	932	1145
Soviet Union[2]	215 (1958)	380	485
Japan	28	167	325
West Germany	50	142	185
People's Republic of China[3]	46	56	71

[1] The projected annual growth rate for Japan used by the Japanese Economic Council in their Economic-Social Development Plan for 1970-1975 is 13% for 1970 and 10.6% for 1971-1975. The annual growth rates assumed for the United States, the Soviet Union, and West Germany were 3.4%, 4.5%, and 4.5% respectively, figures on which there is general consensus among economists. For communist China a liberal 4% rate has been assumed.

[2] In view of the varied ways of computing the GNP of the Soviet Union, the data cited have been derived from a single source: Stanley H. Cohen, "Soviet Growth Retardation: Trends in Resource Availability and Efficiency," *New Directions in the Soviet Economy* (1966), a collection of studies proposed for the Subcommittee on Foreign Economic Policy of the Joint Economic Committee of the Congress of the United States, pp. 99-132. The figure for 1969 is a projection from the 1964 figure given in the report, assuming an average growth rate of 5.3%, which was the average actual rate of growth for the Soviet economy from 1958 to 1964.

[3] Estimates of the GNP of communist China are extremely hazardous and the figures quoted are rough approximations. The 1969 figure is a projection using a generous 4% growth rate from a figure for the year 1966 cited by Dwight Perkins, "The Chinese Economy and Its International Impact," *SAIS Review*, Winter 1968, p. 38.

Other sources: Bruce M. Russett, et al., *World Handbook of Political and Social Indicators* (New Haven: Yale University Press, 1964), pp. 149-57; *International Financial Statistics*, March 1970; and *The Japan Times Weekly*, April 18, 1970.

polarity will bring changes in Japanese political and military foreign policies is related to the evolution of the American alliance and to pressures generated by expanded Japanese involvement in the Asian region. Japan will be able to afford the costs, both materially and technically, to work for superpower military status, but its late start and the limited incentive for such a role make most improbable any effort beyond a concern for regional security problems. Within the next decade, at least, the vastly superior military capabilities and concomitant global political commitments of the United States and the Soviet Union will render even a nuclear Japan a second-level power in international politics except in East Asia.

Because Japan's remarkable economic growth has been tied so closely to expanding international trade, the need to maintain positive connections with the global economy provides an important constraint on any autonomous military or economic action. Although Japan's dependence on imports (7-9 percent of GNP) is well below that of almost all West European nations and a successful effort has been made to diversify the sources of imports, the country still relies heavily on outside supplies of basic raw materials such as oil and iron ore.[7] Similarly, Japan is statistically less dependent on exports (8-9 percent of GNP) than any other country except the United States, but to sustain the momentum of economic growth sought by the Japanese government, it will be necessary to expand exports steadily (at roughly 16 percent per annum) and to take a progressively larger proportion of total world trade. These goals can be reached only by carefully nurturing economic and political ties with the United States in particular and all advanced industrial countries generally. The charges of Japanese economic nationalism now recurrent

[7] Most of the statistics in this paragraph were taken from the report published in 1970 by the Morgan Guaranty Trust Company, "Japan's International Economy."

in both the United States and in East Asian countries would be aggravated substantially by a sudden or rapid expansion of political and military actions on behalf of "national interest." There are external economic disincentives to a policy requiring extensive conventional or nuclear rearmament.

Since the end of the Occupation, Japan has sought to eradicate the stigma connected with the outbreak and conduct of the Greater East Asian war in order, as Prime Minister Sato put it, "to occupy an honored place in international society."[8] This quest for status around the banner of peace has served as the articulated leitmotif of postwar Japanese foreign policy. It has taken on concrete meaning in terms of participation in the United Nations, sponsoring international extravaganzas such as the 1964 Olympics and 1970 Expo, and legitimatizing cautious efforts at aid and cooperation in terms of brotherhood and peace. Doubtless such moral blandishments will not carry much weight when matters of national security are involved (witness the equivocation on the Nuclear Non-Proliferation Treaty), but much more than public relations underlies this quest for international status. Sufficient momentum has been gained by this policy that any reversal would exact rather high external as well as internal political costs.

On the whole, ambivalent influences flow from Japan's participation in the global international system. Both the moral tone of past policies and the web of economic interdependence that have grown up during the past two decades place important limits on any sudden reversal, during the next five years, toward a security posture involving greatly expanded armaments. At the same time, new uncertainties surrounding the American alliance and the complexities attendant on the trend toward multipolarity have placed Japan's international role in the global system in an entirely new light. So have

[8] Speech before the United Nations General Assembly, October 21, 1970, *The Japan Times*, October 22, 1970.

recent developments in the Asian regional system of which Japan is a part.[9]

Beset during the last twenty-five years by two major international wars, the prolonged agonies of decolonization, and continuous reverberations from the cataclysm of the Chinese revolution, this Asian region is and will remain a cockpit for international conflict. From it, Japan can no longer remain aloof. To understand the regional security issues that may confront Japan in the immediate future, it is necessary to appreciate the range of Japanese involvement here.

Economically, Japan is the overwhelmingly preeminent power in the region to a degree that is not usually appreciated. Using GNP as an index of international power (Table 2), Japan is not only three times the size of the People's Republic of China but is seventeen times as "powerful" as the third ranking state (Indonesia) and has more than one and a half times the total GNP of all regional states—including China. The relative advantage of Japan will become still greater in the next four years. Assuming generous rates of economic growth for the other countries in the region (10 percent per annum for South Korea, Thailand, and Nationalist China, and 5 percent for the People's Republic of China and all other countries), by 1975 Japan will have more than two and one-half times the *total* GNP of all other countries in the region. This is a position of superiority greater than that enjoyed by the United States relative to other advanced nations during the heyday of bipolarity in the 1950's.[10] Simi-

[9] This Asian system includes China, Japan, Korea, and Taiwan, together with what is normally considered to be Southeast Asia: Malaysia, Singapore, Thailand, Cambodia, Laos, North and South Vietnam, Burma, Indonesia, and the Philippines. The criteria used in establishing these boundaries include recent patterns of political, social, and military interaction, and the capabilities and policy priorities of the two major actors in the region, China and Japan.

[10] For comparison see Bruce M. Russett, *Trends in World Politics* (New York: Macmillan, 1965), pp. 4-8.

larly, as seen in Table 2, the structure of the East Asia regional subsystem is bipolar in much the way that the global system was in that earlier period. What makes this structural feature important is that at this time, when the differences and antagonisms between communist and non-communist nations in Europe seem to be lessening, the division has become more intense in East Asia, fueled by the uncertainty of Chinese posture and the continued partition of several countries. Thus, the region is divided along political lines, and within

TABLE 2

GROSS NATIONAL PRODUCT OF EAST ASIAN COUNTRIES (1969)
(In billions of U.S. dollars)

Country	*GNP*[1]
Japan	167.4
Communist China	56.2
Indonesia	11.2
Philippines	8.0
South Korea	7.0
Thailand	6.0
Nationalist China	4.8
Malaysia	3.6
South Vietnam	3.0
Burma	2.0
North Vietnam	1.4
Cambodia	1.0
North Korea	1.0
Singapore	1.0
Laos	0.2

[1] The figures for South Vietnam, Burma, North Vietnam, Cambodia, North Korea, Singapore, and Laos are estimates. Communist China's GNP has been estimated by projecting a generous 4% growth rate from a figure for the year 1966 cited by Dwight Perkins in "The Chinese Economy and Its International Impact," *SAIS Review*, Winter 1968, p. 38.

Sources: International Monetary Fund, *International Financial Statistics*, October 1969, June 1970; Agency for International Development, *Selected Economic Data for the Less Developed Countries*, June 1969; Agency for International Development, *Gross National Product: Growth Rates and Trend Data*, February 1971; Agency for International Development, *Economic Data Book for East Asia*, 1968; Bureau of International Commerce, *Foreign Economic Trends and Their Implications for the U.S.*, 1969, 1970.

each grouping just one country, China or Japan, stands with the capacity to act independently and effectively on a region-wide basis on several levels of international politics. The broad patterns of international interaction within the region accent the divisions apparent in this broad structural profile.

The magnitude and direction of East Asia trade development offer a particularly salutary perspective from which to appraise Japan's involvement in regional international relations. From 1958 through 1968 the Japanese increased their share of regional trade from 9 to 27.4 percent, displacing the former colonial powers, completely overshadowing the People's Republic of China, and widely surpassing the United States.[11] Although the U.S. share (despite no trade with Peking) did rise markedly from 12.9 to 21.9 percent, the major portion of this growth came in the years 1965-1968 and largely reflected the very substantial and temporary expenditures related to the conduct of the Vietnam war. Japanese dominance is further reflected in the bilateral trade patterns between Asian nations and Japan. As seen in Table 3, Japan is now the first or second leading trading partner of every nation in the region, a degree of regional ascendancy comparable to that of the United States in Latin America. Clearly, the trade dependence of the nations in the region on Japan is growing, and Japanese trade is now involved in such a large proportion of the international transactions of most East Asian countries that any sharp alteration of these relationships will result in severe economic dislocation. Two other aspects of this trade are of particular significance. Although trade with East Asia has continued over the last decade to take around 19 percent of total Japanese trade, Japan is not in a basic sense dependent on the region, and the asymmetry of relations with all of the individual countries provides Tokyo with the same kind of policy leverage that was earlier

[11] International Monetary Fund, *Direction of Trade*, Annual 1958-1962; Annual 1963-1967; May 1968; March 1969; June 1969.

TABLE 3

TRADE OF EAST ASIAN COUNTRIES WITH JAPAN AS A PERCENTAGE OF THEIR TOTAL TRADE (1968)

(In millions of U.S. dollars)

Country[1]	EXPORTS[2] Amount	%	IMPORTS[2] Amount	%	TOTAL Amount	%	Rank (Total trade)
Burma	12.4	15.2	39.3	24.2	51.7	21.3	1
Cambodia	6.6	21.5	20.3	28.0	26.9	26.1	2
Communist China	224.2	16.9	325.5	29.5	549.7	22.6	1
Nationalist China	150.7	19.1	471.7	52.7	622.4	36.9	2[3]
Indonesia	251.9	34.7	146.6	22.1	398.5	28.7	1
South Korea	101.6	22.1	602.7	46.1	704.3	36.5	1 (identical with U.S.)
Malaysia	343.4	30.6	104.5	15.9	447.9	25.2	1
Philippines	398.0	48.3	411.1	32.1	809.1	38.5	2[3]
Singapore	61.8	20.6	209.3	21.2	271.1	21.1	1
Thailand	147.0	34.4	365.5	35.8	512.5	35.4	1
South Vietnam	2.7	15.4	199.0	28.1	201.7	27.7	2

[1] North Vietnam and North Korea have been omitted because trade data with communist nations is incomplete. Laos has been omitted because it is not statistically significant.

[2] As reported by Japan.

[3] Preliminary reports indicated that Japan had become the leading trading partner of both the Philippines and Nationalist China in 1970.

Sources: International Monetary Fund and International Bank for Reconstruction and Development, *Direction of Trade Annual 1963-1967*; *Direction of Trade*, June 1968, August 1968, February 1969, March 1969, April 1969, May 1969, June 1969, July 1969, September 1969, November 1969, February 1970; *Far Eastern Economic Review Yearbook*, 1969.

noted regarding U.S. trade ties with Japan. Further, Japan's trade with the leading non-communist nations (notably Indonesia, the Philippines, Thailand, Malaysia, the Republic of Korea, and the Republic of China) is conspicuously large and will grow most rapidly if the various schemes to promote regional economic growth bear fruit. In consequence, Japan not only has a strengthened position vis-à-vis other East Asian nations, but this markedly increased economic stake makes involvement in the politics of the region much more probable.

Beyond the network of trade ties generated largely by the forces of the economic market, the international climate has been particularly salutary for the modest moves undertaken by Japan toward an expanded role in the region. Largely economically oriented, multilateral organizations, such as the Asian Development Bank, have sprouted among non-communist Asian nations behind the smoke of Vietnam. Participation has cost the Japanese very little and has yielded not only substantial good will but promoted conditions of economic stability from which Japan stands to profit handsomely. All of Japanese wartime reparations and the overwhelming proportion of its foreign aid are concentrated in the region. Today Japan extends considerably more aid to East Asian countries outside of Vietnam than does the United States and has pledged to double this within the next five years.[12]

The combined effect of these individually modest commitments is to increase greatly the non-military obligations of Japan toward non-communist Asian states. At the same time, it draws Japan inevitably deeper into their politics and security problems, for a number of forces converge to produce deep instability in the region. These forces include: the highly uncertain implications for international behavior of the pro-

[12] Japan, Ministry of Foreign Affairs, *Highlights of Japan's Foreign Aid* (Tokyo, 1969), *passim.*

found economic and social changes occurring in nations other than Japan; the conflicts of interest tied to past and current national competition that are aggravated by ideological cleavages between communist and non-communist political groups; the virtual impossibility of a "power balance," given the permeability of many states to "peoples' wars" and the intransigence of China to any attempt to freeze the status quo; the failure of both superpowers to articulate clearly their short-term objectives in Asia, far less to build effective alliance systems.

In the context of the trends toward global multipolarity and increased Japanese involvement in the highly unstable Asian region, it seems clear that Japan will be confronted by several major security problems during the next five years. Japan cannot participate fully or for long in regional politics with a military role narrowly restricted to the home islands.[13] When and how an expanded military role may emerge will depend on developments regarding four issues: security questions surrounding resolution of the war in Indochina, the China problem, relations with the two Koreas, and relations with the Soviet Union.

The circumstances attendant on the settlement of the Indochina war will provide the most immediate pressures for altering Japanese security policy. Three basic external sources of influence are the United States, the non-communist countries in the region, and expanded overseas Japanese material interests. The character of the U.S. influence will depend on the short-term tactics used in military disengagement from Vietnam and the long-term objective of involving Japan more fully in Asian politics as at least a partial surrogate for the United States. This is well illustrated in two recent policy ac-

[13] The reader may wish to compare this view with that of Kenneth Young, who argues in Chapter VI that Japan can and probably will increase its participation in the politics of the region, even in its security, without playing a military role.

tions. As an addendum to President Nixon's proposal for an immediate ceasefire throughout Indochina to be overseen by an international supervisory team, Secretary of State William Rogers publicly stated that Japan, as well as Malaysia and Indonesia, had "positively expressed interest in joining such a team." Whatever private initiatives were undertaken by Japan on this matter, it is clearly in the interest of the United States to promote this line of action. Similarly, in the conference between President Nixon and Prime Minister Sato in late October 1970, the United States secured a pledge from Japan for a "forward looking posture" regarding greatly augmented aid to the Saigon government.[14] Because aid of this magnitude obviously carries with it political implications, no matter what the expressed motivations of the donor,[15] Japan will be drawn further into the fluid and militarily explosive Vietnam situation. Pressures of this sort will continue until a successful formula for implementing United States disengagement is found. Should the unwinding of the war be protracted, Japan's capacity to remain entirely on the sidelines, even regarding token participation in a military peace-keeping force, will be severely tested.

Strong incentives for further Japanese involvement in the *realpolitik* of Southeast Asia will grow out of the web of regional political and economic interconnections previously discussed. In the probable event that the Indochina conflict becomes "de-Americanized" with a cloud of doubt still hanging over the security of the nations in the area, the substantial Japanese trade and investment in this region would be jeopardized in a way that would at least make a decision not to act an agonizing one. Even more important are the cumula-

[14] *Asahi shimbun*, October 26, 1970. The Japanese subsequently proposed $140 million aid for the governments of South Vietnam, Cambodia, and Laos. *Mainichi shimbun*, January 20, 1971.

[15] On the political implications of economic aid see David A. Baldwin, "Foreign Aid, Intervention, and Influence," *World Politics*, Vol. 21, No. 3, April 1969, pp. 425-37.

tive effects of intensified political contacts of recent years. These were clearly displayed in the Djakarta conference convened after the American intervention in Cambodia. Japan could not avoid playing a major role in this meeting as a "leading Asian nation," and any future conference convened to deal with regional security will require Japanese participation. What would happen if such a group were to agree on a positive multilateral military action is far from clear. To "opt out" of all such actions will cause the Japanese very substantial embarrassment and problems in their relations with Asian nations. But to abide by a decision of this sort would lead to military involvement in Asia at a time and place not chosen by the Japanese government. In any event, the persistence of conflict in and around Indochina will lead the neighboring non-communist states actively to seek a new security arrangement, thereby raising pressures for Japan's military involvement in an area in which policymakers in Tokyo would not *choose* to be involved.

Past Japanese policy has been sufficiently ambiguous to allow future flexibility regarding the Hanoi government should the war be resolved in the communists' favor. Relations with South Vietnam have been restrained and proper. Japan has moved with vigor in the area of trade and plans for substantial economic involvement once the war has ended have been made public, but beyond a hydroelectric plant built as part of a reparations agreement, aid has been token and almost entirely on the "humanitarian" level. Although formally supporting the U.S. position, the Japanese government was at best lukewarm regarding bombing of the north and has made evident in various ways (e.g., a report by a quasi-official inspection mission headed by an ex-career diplomat) that the United States has given disproportionate emphasis to a clearly intractable and essentially nationalist and internal war. At the same time, excepting Russia and China, Japan has become the number one trading partner of the Democratic People's Republic of Vietnam. Thus the Japanese have sought to insulate themselves from the international pressures that will immediately flow from resolution of the current conflict.

Whoever wins the war, a major shift in Japan's security role in the region will ensue only in the face of a clear and present threat to Southeast Asia generally or as a result of fresh and stronger doubts about the U.S. military commitment throughout the region. Even contingencies of this sort, implying a full realignment of the East Asian power balance, are not likely to precipitate a dramatic shift in Japanese security policy before 1975 and will of necessity be integrally tied to the China problem, the major immediate foreign policy issue confronting Japan.[16]

China has bedeviled Japan's diplomacy throughout the twentieth century and continues to hold a unique and crucial place during this transitional period for international politics in Asia. To a degree exceeded only by the United States, China has been central to the foreign policy debate within Japan. The issue transcends party lines and has brought together as political allies a motley group of nostalgic, conservative Sinophiles from the prewar era, opportunistic businessmen in search of the legendary China market, and left-wing Maoist revolutionaries.[17] Since 1964 most conservative dietmen have belonged to the Afro-Asia Study Group or the Asia Study Group, loose associations whose main *raison d'être* is to promote early normalization of relations with Peking (the "A-A" group) or to support the current policy recognizing Nationalist China (the "A" group). Since 1965 a "pro-

[16] For further discussion of how various settlements of the Vietnam problem may affect Japan's relations with other Southeast Asian states, see Kenneth Young's analysis in Chapter VI, pp. 186-88.

[17] See, for example, Donald C. Hellmann, "Japan's Relations with Communist China," *Asian Survey*, Vol. 4, No. 10, October 1964, pp. 1,085-92; and *Asahi shimbun*, November 14, December 8, 16, 1970.

Peking" faction has controlled the Socialist party and in obsessive fashion has used China policy in intra- and inter-party politics and to promote anti-Americanism. When in late 1970 the international tide suddenly turned in favor of Peking's admission to the United Nations, the importance that the China issue holds in domestic politics was abundantly manifested in the complicated moves and countermoves that immediately took place throughout the political world. Adjustments in the external relations involving Japan and the two Chinas are greatly complicated by the salience and definition that the issues already have among politically articulate groups in Japan.

Japan's policy toward China has been conducted on two distinct levels. With the right hand, the Japanese have followed the hard United States line of non-recognition, establishing formal diplomatic ties solely with the Nationalist government, sponsoring in the General Assembly the "important question" resolution that has inhibited Peking's admission to the United Nations, and cultivating extensive economic ties with Taiwan. With the left hand, Japan has, through various cultural, political, and economic missions, established more varied and extensive contact with the mainland than any other non-communist nation and since 1965 has been far and away China's leading trading partner. The result has been a kind of de facto "two Chinas" policy, involving an ostrich-like position regarding the uncomfortable political and security issues raised by the growing international power of Peking and a tendency to define the China problem narrowly in terms of Japan's relations with the two Chinese governments.

The dilemmas found in the triangular relationship between Tokyo, Taipei, and Peking graphically illustrate the inadequacies and weaknesses of the trading company philosophy underlying Japanese foreign policy, that is, to separate economics from politics and abandon *realpolitik*. At first glance, the opposite seems to be true. Although the Nationalists and the communists adamantly insist that there is only one China and that politics and economics are inseparable and have sporadically imposed retaliatory restrictions against Japanese trade, Japan has steadily expanded economic relations with both countries in relative and absolute terms. In 1970 Japan absorbed roughly 40 percent ($951 million) of the total trade of Taiwan, supplying an astonishing almost 50 percent of the island's imports. Additionally, the government has extended $150 million in special yen credits to the nationalists, and Japanese private investment has steadily expanded now to between $80 and $100 million. In consequence, Japan's economic stake, which roughly equals that of the United States, has achieved a size and momentum that has political significance for Tokyo as well as Taiwan. Trade with the mainland will exceed $820 million in 1970; but even more important to Japanese traders is the attraction of the vast potential of the Chinese market, a will-o'-the-wisp that has similarly tempted Western merchants since the early nineteenth century. Communist efforts to link politics with trade in the past have been largely ineffectual (since 1964 there has been almost a direct correlation between the political invective heaped on the Japanese government and an *increase* in trade), and in purely economic terms the advantage will continue to rest with Japan. However, the spate of moves among Western industrial countries in late 1970 to recognize Peking has heightened the concern of the Japanese business world regarding future competition and forced the government to face squarely a major political choice likely to involve substantial economic cost no matter what policy is adopted. Moreover, the political alternatives involved are both complex and risky.

Japan is bound to the Republic of China on Taiwan by far more than economics. In 1952, as one of the American-stipulated conditions for the San Francisco peace treaty ending World War II, a separate peace treaty was signed with Chiang Kai-shek's government. For the Japanese this latter treaty ended one of the ugliest aspects of the war (for which they felt a particular moral onus), publicly established an obligation to support the Taipei government internationally, and specifically exonerated them from any reparations obligation. Another special and important link with Taiwan grows out of the legacy of the fifty years of Japanese colonial rule over the island. This has been most visibly manifested in the attitudes of several leaders of the "Taiwan lobby" within the Liberal Democratic party, who have successfully promoted closer relations between the two countries.[18] It also underlies more broadly shared sentiments that the Taiwanese should be provided with at least the choice of whether or not they wish to join the communist mainland. Finally, in the November 1969 communique announcing the American return of the Ryukyu Islands, Japan explicitly acknowledged a special security interest in Taiwan. A quid pro quo demanded by the Americans, this general commitment took on added importance during the ensuing year as the plans unfolded for immediate and substantial reduction of American troops deployed in Asia and as the Nixon Doctrine raised serious doubts about ultimate United States commitments in the region. Japan could not, even if it would, at this time make a meaningful contribution to Taiwan's security, but as a possible harbinger of the future and as an obstacle to normalized relations with the mainland this statement carries considerable importance. It further strengthens the pattern of interaction between Japan and non-communist Asian nations, is consonant with the American goal of an Asian defense system under Japanese leadership, and aggravates the fears of the Chinese communists that Japan will indeed play a major military role in the region. In view of these circum-

[18] Interview with Mitsujiro Ishii, Liberal Democratic party faction leader, May 29, 1962.

stances, a Japanese move to recognize Peking would almost inevitably involve a "two Chinas" formula to salvage the political and economic interests of Taiwan.

Yet there are strong reasons for Japan to move boldly and rapidly toward rapprochement with the Chinese communists. Every prime minister has recognized that Japan's policy toward the People's Republic is the cornerstone for relations with all Asian nations, a position rooted in the realities of the regional conditions of international power as well as in the special attraction China continues to hold for the Japanese. Up to the present Japan has not taken an unambiguous and independent stand, choosing instead to remain in the American shadow on political matters while aggressively pursuing trade. From an overall perspective Japan's China policy has been schizophrenic. Regarding Peking, it has been prayerfully passive. Bilateral relations have been determined almost entirely by initiatives from the Chinese or the drift of world politics. China's criticisms of Japan have been monotonously consistent and have concentrated on opposition to the alliance with "American imperialism," to collaboration with the "Taiwan bandits" and the "American puppet regime in South Korea," and, more recently, to the rise of "Japanese militarism." What is likely to initiate change in Tokyo-Peking relations is not a policy shift by the Chinese, but the now evident alteration in the climate of world opinion to bring China more fully into international society and the pressure this trend places on Japan to acquiesce.

Few advantages are likely to flow from normalization of relations, and the actual pattern of intercourse should not alter greatly in the short run. Certainly a dramatic spurt in economic activity is improbable, both because of limitations of the economic market in China and because long-term contracts in extractive industries (e.g., coal and iron ore), in which Japan is interested, presuppose a level of political confidence and stability that will not emerge in the immediate future. The very idea that Japanese diplomacy, with its record of equivocation and stolidity, will provide a unique bridge for communication and/or influence over the committed and strong-willed leaders of the greatest revolution in modern history beggars the imagination. Ultimately, the question is how recognition of Peking will affect the international politics of East Asia and thus the security of Japan. The answer must be, very little, even though some capacity to communicate is obviously better than none. International stability and peace in Asia will flow from a complicated set of forces, basic to which will be some sort of military and political "power balance." It is gratuitous to assume that following the reduction of the American military presence, the Chinese will withdraw from all efforts to employ *realpolitik* to expand their international influence because of "logistical weaknesses," the enormity of internal problems, or the like.[19] China is in the throes of revolution under a leadership militantly nationalist, holding universalist and radical ideological commitments, and cultivating sympathetic or allied revolutionary communist groups in every underdeveloped Asian nation. Whatever the short-term effects of recognition might be, it seems possible, indeed likely, that the Chinese will ultimately stand as international competitors of the status quo-oriented Japanese, quite independent of the level of diplomatic intercourse. Sino-Japanese competition in East Asia will necessarily involve matters of *realpolitik*, and perhaps the most plausible impetus for seeking some accommodation with Peking at this time is to forestall a collision course that could precipitate an untimely and costly expansion of security capabilities.

More than any single influence, relations with the People's

[19] For elaboration of a contrary position see "The Aggressive People's Republic of China: Menace or Myth" (a colloquy among Edwin O. Reischauer, J. W. Fulbright, Mark O. Hatfield, and Ezaki Masumi), in Elaine H. Burnell, ed., *Asian Dilemma: United States, Japan and China* (Tokyo: Charles E. Tuttle Co., 1969), pp. 81-90.

Republic of China will determine how Japan approaches the question of rearmament during the next five years. What possible security threat can the Chinese pose for Japan? If the Japanese continue to limit their security zone solely to their home islands and accept in toto the American nuclear guarantee, there is not likely to be a clear or present military danger to them from China, or anyone, that could not be met by their own modest conventional defense forces. Yet the Japanese government has itself acknowledged since 1965 that China (not Russia) does pose a security threat to Japan, and that this threat has a nuclear dimension. Moreover, a Chinese nuclear attack would certainly not be directed *in vacuo* against Japan but would invoke Sino-Japanese conflict over an "outside issue" or the conditions of alliance with the United States. That is, it would most likely take the form of nuclear blackmail. Thus a China threat is tied to Japanese engagement in Asia, either through autonomous commitments involving security considerations or from alliance with the United States. The latter alternative is increasingly less probable in view of gradual U.S. disentanglement from Asia. A Japan engaged in East Asia could avoid a real or perceived threat from China only in the highly improbable event that international conflict is purged from the region or the Chinese effectively withdraw from international politics.

Participation in power politics in Asia with China as its main rival will put enormous pressure on Japan to develop an independent nuclear force. The timing of any such move is closely tied to the continuing credibility of the U.S. nuclear umbrella and Chinese acquisition of a delivery system with capabilities to threaten the continental United States. Because the momentum of U.S. policy is toward military disengagement from the region, convincing Tokyo that our nuclear intentions are unequivocal will prove progressively more difficult. Whether Japan embarks on a path toward nuclear armament will depend not so much on the designs of political leaders in Tokyo as on the ways in which the two dominant regional powers are caught up in the maelstrom of international politics in East Asia.

From this perspective, U.S. insistence on Japanese ratification of the Nuclear Non-Proliferation Treaty is risky and dubious policy. This has raised the one security issue on which Japan may be forced to break openly with the United States—an alternative clearly implicit in the long and sharply phrased statement issued by the Foreign Ministry when Japan signed the treaty. Emphasizing the inherent inequities for the non-nuclear powers, the government made clear that ratification by Japan was contingent, among other things, on "concrete measures of nuclear disarmament . . . by *all* nuclear powers." China, a non-signatory of the treaty, was also specifically singled out as a country of special concern. In the first postwar White Paper on defense issued in October 1970, the Japanese government further stated that "tactical" nuclear weapons could be developed under the present constitution. The nuclear option, in short, will be kept open, and variables conducive to a move in this direction are almost wholly outside the control of the Japanese decision-makers.

Korea, a divided country designated with Taiwan as an area of special security interest for Japan in the November 1969 Sato-Nixon communique, provides the most immediate potential setting for a military conflict involving both the Japanese and the Chinese. In many ways Japan's relations with the Republic of Korea (South Korea) are similar to those with Taiwan. Since diplomatic normalization in 1965, economic links between the two countries have expanded at a remarkable rate. Japan absorbs almost 40 percent of the total ROK trade, providing nearly 50 percent of imports. At the end of 1969 more than $140 million of Japanese capital had been or was to be invested in the south and commercial loans in that year alone approximated $100 million.[20] Again, both the size of and the momentum behind this material stake in the Republic of Korea have political significance. The "Korean lobby" within the conservative party is less powerful than that supporting Taipei, but it involves many of the same individuals, including former Premier Nobusuke Kishi with strong anti-communist sentiments and a nostalgia for the time when Korea too was part of the Japanese empire. In sharp contrast to the mystique that surrounds China, Korea and Koreans are generally disliked and looked down upon by all strata of Japanese society. A reciprocal bitterness is felt by the Koreans over the years of harsh colonial rule. Any bilateral military arrangement would be born for almost purely expediential reasons, but the persistence of the Korean security problem in the face of continued uncertainties regarding American military commitments and an increasing web of interdependence between Japan and the south could raise in clear and specific terms the pressures for an autonomous military role for the Japanese in post-Vietnam Asia.

A major check on the confrontation in Korea has been the deep involvement of the United States and the Soviet Union with the respective sides. The resulting Cold War type of confrontation has brought the usual constraints found when the superpowers directly face one another. This situation is changing. Even if the United States remains ultimately committed to defend Korea, after Vietnam it is very uncertain whether a full scale "limited war" in the style of the 1950's would or could be undertaken. Furthermore, even if the Soviets were prompted by global strategic considerations to forestall or limit any outbreak of hostilities, it is far from clear that the Chinese, with narrower strategic interests and a more immediate concern for the military importance of Korea,

[20] The figures were taken from two 1970 publications of the Economic Planning Board of the Government of the Republic of Korea, *Major Economic Indicators* and *Economic Survey*.

would be so constrained. In any open conflict of this sort, the critical pawn in the ensuing strategic game would be Japan, a nation essential to the effective conduct of a conventional war by the United States and immediately susceptible to a nuclear threat from China. Any Chinese effort to neutralize Japan in this way would test the credibility of the U.S. nuclear umbrella and, whatever the result, bring Japan face to face with the *realpolitik* of East Asia. The whole issue of rearmament would thus be cast in a very different light.

Following the normalization of diplomatic ties in 1956, bilateral relations with the Soviet Union have undergone a substantial transformation, especially in economic terms. Trade has risen from almost nil to total more than $800 million and tentative agreements have been reached for extensive Japanese participation in joint development projects in Siberia and the Soviet Far East. Moreover, the potential for future growth is great. Russia could easily become an even more important source of raw materials (e.g., oil, lumber, and natural gas) needed by Japan, and there are strong incentives for broadened Japanese economic activities in accelerating the development of the eastern territories. Non-economic ties have also markedly improved as reciprocal visits by government leaders and frequent trips by Japanese businessmen have brought into being a level of communication and interaction exceeding all past expectations. Despite these positive achievements, however, Soviet-Japanese relations are still essentially coldly formal, beset by unsolved bilateral problems and by mutual uncertainty regarding their future respective roles in East Asia.

Two factors in Japanese domestic politics are particularly important for Soviet relations: the long-standing negative attitude of the Japanese toward Russia and the still-outstanding territorial dispute regarding ownership of the Southern Kurile Islands. Throughout Japan's modern history, Russia has been seen as a menacing neighbor, a challenge to the nation's security if not an enemy in war. Soviet actions since 1945 have done little to improve this image. Moscow unilaterally broke a non-aggression treaty to participate in the last days of the Pacific war and incarcerated for several years thousands of Japanese civilians and military personnel captured in Manchuria and North Korea. Subsequently, Japanese fishing in the northern seas near the Soviet coast has been restricted and harassed, and all territorial claims have been rejected. From the time of the Korean war the Russians have also loomed as a protagonist of the Cold War and a threat to the peace and security of the country. Understandably, the Soviet Union has consistently ranked as the "most disliked" country in opinion polls throughout the postwar period. Such widespread and intense feelings cannot but inhibit the establishment of truly close relations and augment the opportunities for conflict.

The territorial issue is and is likely to remain a major stumbling block in bilateral relations. Japan's historical claim to the disputed territory is strong. Its legal case is not. Under the terms of the Yalta agreement, the San Francisco peace treaty (Article 2, Section 3), and the Diet testimony of Prime Minister Yoshida, head of the delegation to the peace conference, Japan renounced all claims to these islands.[21] Since August 1955, however, Japan (then in the midst of negotiations for normalization of relations) has adamantly claimed the Southern Kuriles, and failure to settle this issue has prevented a formal treaty ending the war. After fifteen years the campaign of the Japanese government to regain these lands, with its nationalist and irredentist appeal, has gained a momentum of its own that makes very unlikely any early or compromise settlement.

[21] See Japan, Shugiin (House of Representatives), *Heiwa joyaku oyobi Nichi-Bei anzen hosho joyaku tokubetsu iinkai giroku* (Proceedings of the Special Committee on the Peace Treaty and the Japan-United States Security Treaty), No. 4, 12th Diet, pp. 18-19; and Hellmann, *Japanese Domestic Politics and Foreign Policy*, pp. 34-35, 59-60, 146.

From the onset of the Cold War until the mid-1960's the Soviet Union was seen as the only possible security threat to Japan. This assumption underlay both the 1952 and the 1960 security treaties with the United States and most of the planning embodied in the five-year plans under which Japanese military forces have been expanded. However defensible in terms of Cold War logic, the issue is now much more complex, primarily because of the independent and nuclear foreign policy of the People's Republic of China in the shadow of the continuing global stand-off of the superpowers and the emergence of Japan as a major regional power. A conventional invasion of Japan by Soviet troops, a basic consideration in initially expanding the Self-Defense Forces, was never really a plausible assumption and today is utterly unthinkable.[22] In addition to the difficulties for Moscow inherent in reversing its long-standing commitment to the policy of peaceful coexistence, the Soviets would inevitably risk full-scale retaliation by the United States in any direct military assault on Japan. The constraints of bipolar strategic confrontation will apply with or without the U.S.-Japan security treaty.

Basic to understanding the changing nature of the Soviet threat to Japan are the altered strategic and political goals of Moscow in Asia. In the words of the 1950 Sino-Soviet Treaty of Friendship and Alliance, "Japan and any nations allied with it" were the main obstacles to peace in East Asia. Now it is China, not Japan, that the Soviets seek to contain in Asia —an aim shared by the Americans and the Japanese. The tolerant attitude taken by the Soviets toward the extension of the American security treaty in 1970, contrasting sharply with their position in 1960, reflects this change in goals. How to "contain" China in the international quicksands of Asia

[22] For a discussion of the assumptions underlying Japanese-American strategy, see George F. Kennan, "Japan's Security and American Policy," *Foreign Affairs*, Vol. 43, No. 1, October 1964, pp. 14-18.

has confounded the Russians as well as the United States. The Soviet Union has already begun to maneuver to strengthen the now limited and tenuous contacts with non-communist nations on the periphery of China. But the latitude for maneuver regarding these nations and the communist buffer states immediately adjacent to China is very limited. Will the Russians, after Vietnam, risk direct military involvement should this be required? How can they cope with initiatives of the revolutionary, Chinese-oriented communist movements? Like the United States and Japan, the Russians will in the immediate future be acting as essentially a status quo power in a period of profound upheaval in Asia. In one sense, as the United States reduces its involvement, the Soviet presence will increase, and to the extent that the Russians themselves are caught up in the swiftly changing circumstances within the East Asian region, they will become directly involved in the external pressures conducive to change Japan's security policy.

Japan is today the sleeping giant of Asia, indeed of the world, with enormous untapped capacities for action. When it will become a major actor in international politics and whether this will involve nuclear capabilities will be determined fundamentally by external pressures. The situation is thus unpredictable and precarious. There are truly substantial constraints on any decision to rearm in the next few years, but the future ultimately depends on the dynamics of East Asian politics, on variables that are both changing and demonstrably beyond the control of any nation. A *volte face* is doubtful, but for Japan a period has commenced in which decisions will be made shaping the pattern of foreign policy for the next generation. The fate of a bourgeois and status quo nation in a revolutionary setting is, of necessity, hazardous and insecure.

CHAPTER VI

The Involvement in Southeast Asia

KENNETH T. YOUNG

THE years 1969-1970 were a major watershed in contemporary Asian history. The old era of Western supremacy was ended. European colonial power had already disappeared and now the American presence began slowly to recede. A new era of Asian primacy in Asia had begun—an era in which Japan is emerging as *the* preeminent power in the region. How Japan will use that power, therefore, becomes an increasingly critical question. As a Japanese official is reported to have said recently: "We are involved in a serious debate about what role we should play, politically and militarily. Economically we follow a very aggressive policy in Asia. . . . Politically we are just beginning to put our toe in the water as far as playing a larger leadership role is concerned. Militarily, we are very reluctant to do more than build up our own self-defense capacity."[1]

But what of the future? The history of Japan's relations with Southeast Asia offers little guidance. Except for its piracy in the sixteenth century, Japan did not play an economic, political, or military role in the region before modern times. China, then the West, have been the paramount external influences, Japan challenging their dominance only for a brief period during the Pacific war. A weakening Japanese empire in its last days gave a great boost to Southeast Asian independence in 1945; but the war left its scars, and since 1951 Japan's reentry into Southeast Asia has been deliberately slow, careful, and commercial. As a result, during most of the past two decades Japan has not figured prominently in the policies and attitudes of Southeast Asia. The United

[1] *New York Times,* August 15, 1970, p. 8.

States and China have filled the scene. Japan has appeared remote and reticent—a land of commercial profit-makers rather than political risk-takers.

By 1970, however, a turning point had clearly been reached. The United States and China were no longer the major external actors in Southeast Asia. Japan had entered. The legacies of the Pacific war had faded, only to be replaced by a new anxiety over the impact of Japan's enormous economic power and the potential of Japan's political and military resurgence. No Southeast Asians wanted the ill-fated Co-prosperity Sphere to be recreated; and, as has already been pointed out, a recommitment of Japanese military power to that region in the foreseeable future seemed unlikely.[2] But the economic involvement of Japan was growing and it seemed inevitable that Japan's concern for the security of the region would increase. The question of the 1970's was, therefore, barring a major shift in the world balance of power that might provoke a fundamental change in Japanese policy, how and to what extent Japan would participate in the security of the region if not in a directly military way. What would the Japanese be willing to do? What would the Southeast Asians want them to do?

The preponderance that Japan has achieved among the economies of Asia—its GNP nearly three times larger than that of the People's Republic of China or of all of Southeast Asia combined—and the enormity of Japan's share of all Southeast Asian regional trade—27.4 percent in 1968 (including that of the PRC)—have already been pointed out.[3] Even these impressive ratios may be expected to increase. For 1975, for example, some Japanese economists predict that

[2] See Donald C. Hellmann's conclusions on this subject in Chapter V, pp. 153 ff.; but note Hellmann's differences with the thesis of this chapter that Japan's involvement in Asian political and security matters may be deepened without becoming military.

[3] See Donald C. Hellmann's discussion in Chapter V, pp. 148 ff.

Japan will be supplying as much as 35 percent of total Southeast Asian imports, a doubling since 1965.[4]

In the period from 1965 to 1970 Indonesia and the Philippines seemed to be the foremost commercial interests of Japan, although Thailand followed closely.[5] In fact, Indonesia seems to be emerging as Japan's first economic priority in Southeast Asia. Tokyo organized the first conference of a consortium to aid Indonesia in 1966. The Bank of Tokyo opened a branch in Djakarta in 1968, while nearly one hundred companies had representatives there. With Indonesian encouragement the Japanese moved vigorously into production-sharing or purchasing ventures in oil, timber, nickel, bauxite, and rubber. Special new firms opened up, such as Japex Indonesia for oil exploration. A high-level Japanese business mission went to Indonesia in 1968 but left with uncertain results. In Indonesia as elsewhere in Southeast Asia, the Japanese showed inadequate interest in Indonesia's development and general ignorance or indifference about Indonesian customs and history. Indonesians feared Japanese exploitation; Japanese feared Indonesian instability.

In the Philippines Japan became second to the United States in exports and imports during 1965-1970. Here too the Japanese focused on natural resources such as timber and sugar. The products spread throughout the islands. Japanese investment appeared to be relatively high, perhaps between one and two billion dollars—American investment being about $1.5 billion. Nearly ten thousand Japanese businessmen congregated in Manila. Yet, despite or because of these

[4] Saburo Okita, "Japanese Economic Cooperation in Asia in the 1970's," in Gerald L. Curtis, ed., *Japanese-American Relations in the 1970's* (Washington, D.C.: Columbia Books, for the American Assembly, 1970), pp. 110-12.

[5] This discussion of Japanese commercial activities relies heavily on Lawrence Olson, *Japan in Post-War Asia* (New York: Frederick A. Praeger, 1970), pp. 174-223.

economic relations, misunderstandings and perplexities grew on both sides.

In Singapore Japanese commercial activities also expanded with remarkable velocity. By 1969 Japan's share of total investment in Singapore reached 10 percent, almost equal to that of the United Kingdom. The trade balance favored Japan by three to one. Its most important enterprise in Singapore, and perhaps in Southeast Asia, was the Jurong Shipyard, a joint enterprise with the Singapore government. Capable of taking 100,000 tonners, it will make Singapore the major center for repairing and servicing ships between the Persian Gulf and Yokohama. This is a long-range project directly involving Japan in the future security and stability of Singapore and its vicinity.

In Malaysia the Japanese Malayawata steel mill established a long-range base in the Malaysian economy and a projection of Japanese influence in Southeast Asia. Several hundred Japanese representatives were sent to Malaysia concerned with over thirty joint ventures, including, for example, a tin smelter, a blanket factory, a home appliance assembly plant, a food seasoning plant, an advance cable plant, a cement, and a toothpaste company. All in all, Japanese equity investment amounted to over ten percent and was moving up. The Japanese also were constructing two dams in north Malaysia for a World Bank project.

In Thailand it appeared that by late 1970 Japanese investment reached probably more than $200 million, exceeding American, and Japanese imports continued to blanket the country. Over ten thousand Japanese salesmen, promoters, and technicians were centered in Bangkok, making up the largest of the new "overseas Japanese" gatherings in Southeast Asia. Japan seemed to have gained the preeminent commercial status in Thailand by 1970. Six of the ten automotive assembly plants were Japanese-owned or principally Japanese-owned. Japan bought all of Thailand's corn production and seemed to dominate the synthetic textile industry. The Japanese were involved in a steel mill, a glass plant, and a caustic soda plant. Yet Thailand had a three-to-one adverse balance of trade with Japan.

In all these countries the pattern of Japanese activity was identical and the reaction throughout Southeast Asia has been much the same: the Japanese were interested only in short-range projects, focused on resources for Japanese industry, not on local manufacture, and limited their aid to projects connected with the output of Japanese business. The Japanese seemed reluctant to transfer their own resources of capital and know-how to Southeast Asia. They did not "add value" in the economic sense to their investment there. In terms of behavior, most Japanese appeared clannish, self-centered, and "tense and opaque in social relationships."[6]

During 1965-1970 Japanese "aid" to Southeast Asia changed in content, form, and amount. Reparations payments were largely fulfilled by 1970-1971. Government and commercial aid followed a mixed pattern. Although relatively large, most of it was trade-oriented for the benefit of Japanese industry, so that it had less impact on the economies of the recipient than might be expected. Out of a total aid to Asia during 1958-1968 of $2,644 million as defined by the minister of foreign affairs, Indonesia received about $451 million, the Philippines $470 million, Thailand $160 million, Burma $180 million, and South Vietnam $44 million. In addition Japan contributed $17 million to the Mekong project.[7]

In 1969-1970 Japanese leaders began to emphasize that a new stage in Japanese aid would begin that would greatly increase, perhaps even double, the volume of aid to the whole Asian region by 1975. Moreover, it would include more in the form of concessionary aid in capital and technology. Thus, the Japanese foreign minister told the Ministerial Conference for the Development of Southeast Asia in Djakarta in May 1970

[6] *Ibid.*, p. 193.

[7] *Ibid.*, p. 221.

that Japan pledged $4 billion of total aid by 1975.[8] This figure presumably would include the same large portion of short-term export credits and private investment tied to assured procurement of resources for Japanese industry. Yet, a ten-year projection by the Ministry of International Trade and Industry indicates that Japanese economic growth will not only enable Japan to increase its total aid to Southeast Asia substantially during 1970-1975 but will begin to increase the amount of development-oriented portion.

Of course, Japan during 1965-1970 had already demonstrated interest in Southeast Asian development. In 1966 Japan took the initiative in calling the first ministerial conference for the development of Southeast Asia and has since vigorously participated in its annual meetings. Japan also backed the creation of the Asian Development Bank, worked for its location in Japan (which failed), and contributed some $200 million to its $1 billion capital fund. A Japanese was named the first president of the bank.

While all countries have benefitted from this rapidly expanding economic relationship, one unfortunate consequence has been the evolution of rising Asian allergies. The swarm of zealous Japanese salesmen, the flood of Japanese products, and the spread of Japanese investments have, temporarily at least, produced many negative reactions. For example, Thailand's minister for economic affairs has publicly criticized Japanese policies as "high-handed and insensitive." He accused Japan of trying to exploit the weaknesses of Asian countries or even of seeking to dominate them economically. He also made a comment, often heard in the rest of Asia, that the Japanese stay to themselves: "They fly in on Japan Air Lines, are met by Japanese guides, ride to Bangkok in Japanese busses, where they stay in Japanese hotels and eat and drink in Japanese restaurants, all staffed by Japanese. . . ."[9]

[8] *New York Times*, June 22, 1970.

[9] *Boston Globe*, November 25, 1969, p. 2.

The impression unfortunately seems to be growing among many Asian leaders that Japan may be more difficult for them to deal with than will Communist China. In late 1969 the foreign minister of Indonesia, Adam Malik, expressed this growing allergy toward Japanese power as follows: "Japan, through its big and overwhelming economy, cannot be but an object of envy, suspicion and fear among its Asian neighbors, especially since the experience of almost all the Asian countries with Japan during World War II was none too happy.

"The fears against a probable Japanese domination are real. The Japanese have themselves to blame for this. It is the experience of a number of Asian countries that the Japanese still want to treat other Asian countries as the source of raw materials for their industries and as markets for their manufacturing industries."[10]

This stern reproach aptly echoes a widespread reaction to the zeal with which the Japanese go after business and their success in getting it. Southeast Asians are also concerned about the close connection between Japanese private companies and the Japanese government on the grounds that both work closely together for the long-term advantage primarily of Japan. Japanese economic activity in Southeast Asia may be strengthening its economic prospect but it is hurting the psychological outlook for developing mutual understanding and practical cooperation between these countries and Japan in political and military matters as well as economic. Nevertheless, despite the growing fear of Japan's power and resentment over Japan's economic penetration, Southeast Asians generally believe that Japan—unlike India—belongs to the area of East Asia and will inevitably be involved in the future of Southeast Asia. The question is not whether Japan will be involved, but how much and on what terms.

While they have welcomed Japanese economic aid and

[10] *New York Post*, November 24, 1969, p. 40.

trade despite their misgivings, Southeast Asians have usually resisted even the few Japanese political initiatives of the past years. For instance, when the Japanese government sought to have the Asian Development Bank located in Tokyo, the other Asian members insisted that it be put elsewhere. In all Asian groupings, whether at the United Nations or in Asia, the Japanese have had to be careful not to assert the political leadership that is logically related to their economic power.

An exception was the conference of eleven foreign ministers held in Djakarta on May 16-17, 1970, to discuss the situation in Cambodia. Here the Japanese government showed an inclination to become more politically involved in Asian problems, and the non-communist governments of the region extended a welcome. Indonesian Foreign Minister Malik took the initiative along with Japan in proposing the Djakarta conference. While some members apparently considered Japan's suggestions for dealing with the war in Indochina too conciliatory or too general, the conference was nevertheless successful and significant. For one thing, it did include the Japanese, who played an important political part in the proceedings and the follow-up. It was the first time, moreover, that the non-communist governments of East Asia, from Japan to Australia and New Zealand, met together to discuss and adopt recommendations on an Asian security problem without the presence of the United States, Great Britain, France, or the Soviet Union. Japan was one of the three members selected by the conference to follow up its recommendations by holding consultations with the co-chairmen of the Geneva Conference and others to restore peace in Cambodia.

This participation reflects a growing convergence of views between Japanese and Southeast Asians on certain political questions. They have come to share, for example, a similar emphasis on the non-military priorities of political stability, national cohesion, economic growth, and balanced modernization throughout Asia. Military expenditures are regretted.

It is generally recognized that an increase in national or regional forces for security will have to come out of local resources at the expense of putting more into economic and social development. The size and costs of Southeast Asian military establishments are small—outside North and South Vietnam—in relation to population, annual budgets, GNP, and the military power of the United States, USSR, and China. For example, Thailand's security budget has been about 20 percent of its total budget for the last decade. Others are less. Nevertheless, even today Southeast Asian conventional forces divert much-needed internal resources from critically needed development. Any pressure, therefore, to expand local military forces or facilities would be taken as a mistaken diversion from social and economic priorities. Furthermore, forces especially designed to deal with the likely threat to Southeast Asian security—insurgency—would be less costly than the maintenance or expansion of conventional forces if they could be reduced in accordance with the similar assessment from New Delhi to Tokyo of threats to security.

Views on India in Tokyo and Southeast Asian capitals also seem quite similar. India is not expected to assume an important role in Southeast Asia—economically, politically, or militarily—except for its chairmanship of the International Control Commission in Indochina. It is generally believed in East Asia that India will remain too preoccupied with vast internal problems and external threats to be counted on to contribute much assistance or adequate resources to development and security in Southeast Asia or the western Pacific. Proposals for a Tokyo-New Delhi "axis" to shape a new Asian balance of power as a counterweight are not welcome or seriously considered. India's contribution to peace and stability in Asia will come from India's success at home. On the other hand, these views about India's peripheral role could change. The dynamic changes in the Asian environment and some new "surprise" factors could stimulate Indian inclusion in broad political and security interchanges. Nuclear developments could play a key role in this regard.

In addition, Southeast Asian and Japanese leaders share a comparable commitment to the concept and practice of regional cooperation. They continue to use much the same language even if they sometimes mean different things. Thus Foreign Minister Aichi stated his government's policy in an address to the Japanese Diet on February 14, 1970: ". . . the government, while strengthening friendly relations with other Asian countries, wishes to further strengthen the functions of the Ministerial Conference for the Development of Southeast Asia, ASPAC, and so on, in order to enhance interregional solidarity and promote regional cooperation; the government also wishes to strengthen organizations for international cooperation such as ECAFE and the Asian Development Bank in cooperation with the developed countries of other regions, thereby contributing towards the development of the entire Asian region."[11] Officials of Thailand, Indonesia, Australia, New Zealand, the Philippines, Malaysia, Singapore, and other nations in the area have likewise played up the need for regional solidarity including Japan, and by 1970 a framework and pattern of Southeast Asian-Japanese connections had materialized.

This does not extend, however, to the military field. In October 1969 several Japanese vessels did visit Southeast Asian waters and Singapore on a training mission without trouble,[12] but beyond that the military relationship has been nil. The Southeast Asians have asked for nothing and the Japanese have offered nothing.

If present Southeast Asian attitudes are projected without

[11] Kiichi Aichi, *Policy Speeches at the 63rd Session of the National Diet by Prime Minister Eisaku Sato and Foreign Minister Kiichi Aichi, February 14, 1970* (Japan, Ministry of Foreign Affairs), p. 12.

[12] *New York Times,* October 6, 1969.

change to the mid-1970's, it seems likely that the governments of the region will want Japan to expand its trade and aid, but with certain qualifications and conditions to prevent Japanese economic domination. They will increasingly welcome Japan as a major partner in regional organizations and as a participant in regional discussions of political and security issues, but their nationalistic sentiments are too strong for them to want Japan to assume the role of "Asian political leader." Although contingencies can be envisaged where Japanese and Southeast Asian views might converge on the desirability of Japan's cooperating in protecting the sea lanes or in supplying military and related equipment, at first on a reimbursable but possibly later on a soft loan or even grant basis, there is no reason now to suppose that even these limited commitments will be either sought or offered.

Certain dynamic forces could, of course, change this forecast. Increasing economic interdependence is one of these. If trade with Japan does reach 35 percent or 40 percent of Southeast Asia's total exports and imports, it will be vitally important to Southeast Asian economic development. Development is Southeast Asia's goal. The huge and spectacular expansion of the Japanese economy will demand more imports, particularly of raw materials, from Southeast Asia, and increasing investment there for the production of component parts and manufactured goods. Southeast Asia will also be important to Japan as a source of cheap skilled labor for Japanese factories there. In addition to trade, Japanese aid to other Asian countries including Southeast Asia will go on growing in volume and impact.

Concerning economic interchange, Southeast Asians realize the benefits as well as the dangers of this growing interdependence of markets, trade, technological assistance, and capital investment. But they are beginning to insist that the terms of trade and investment be equalized to the extent possible. They want Japan to buy more from them so that exports and imports are balanced on a bilateral basis. They want Japan to invest in local industries in order to generate local employment, income, and markets. In other words, they want Japanese industry and government to show more interest and put more resources into the economic and social development of their individual countries than heretofore.

The general question of Southeast Asian attitudes toward Japan in the political and security tracks is accordingly quite dependent on the extent to which the Japanese quickly and intensively alter their policies and practices to meet those Southeast Asian requirements in economic relations. If the Japanese adopt a policy of devoting more of their enormous economic resources to nation-building and economic development in Southeast Asia, they will help to reduce the rising allergy to Japanese economic penetration. Tactful behavior will also help greatly. These changes could smooth the way for an increasing political role for Japan in Asian affairs and possibly some form of a military role at a later date if needed.

Another dynamic factor could be a growing appreciation of their common concern for the security of sea lanes and air communications. Economic interdependence will strengthen that orientation, as the Jurong Shipyard in Singapore implies. Asia is a littoral continent between two oceans, the Indian and the Pacific. Asia, even with China, is a series of coastal lands from Pakistan to Japan. They look southward and seaward. Southeast Asia is the focal position between South Asia and Northeast Asia connecting these two vast oceans. Some sixty percent of Southeast Asia is water and some sixty to seventy percent of the Southeast Asian people live on islands or on rivers leading into the sea. So do the Japanese. Fish and ships are staples for Southeast Asia and Japan. Most of the modern capitals of Southeast Asia face the sea. So does Tokyo. No major power, Asian or otherwise, and no Southeast Asian country can remain uninterested in the accessibility of the water highway that runs through the Strait of

Malacca and the South China Sea. Above all, it is Japan's lifeline to the Persian Gulf and Europe. Australia also is "a child of the two oceans."[13] Thus the assured stability of and access to this Southeast Asian land and water bridge from Thailand to Australia and from one end of Indonesia to the other may create a vital joint interest for those countries, Japan, and their friends and supporters.

A third dynanic factor that might bring about collaboration between Japan and Southeast Asia for the security of the whole region would be the repercussions of a rapid or substantial American disengagement and withdrawal from the western Pacific. This is not to say that any such major reduction of American power is now contemplated or even likely. Yet what the Japanese or other Asians think and do in anticipation of this contingency will be important during the next five or ten years.

Given a mutual interest in the stability of sea and air communications in Southeast Asia, their growing economic and possible political interdependence, the possibility of American decommitment and disengagement, the expansion of the Soviet presence in littoral Asia, and the continuing hostility of mainland China and North Vietnam toward non-communist governments of Asia, the Southeast Asians and the Japanese might begin to discuss contingency plans for security operations and for military-naval-air supply and logistics to prevent a military vacuum to some extent. For instance, Japan could become a major supplier for coastal patrol ships, electronic communications, and weapons and other equipment for counter-insurgency forces in Southeast Asia. If the threat seemed worse and the possible vacuum more likely—assuming a full American withdrawal from Indochina and Thailand after 1972—Japanese defensive air and naval units might be accepted in Southeast Asia as the

[13] George Thompson, "The New World of Asia," *Foreign Affairs*, Vol. 48, No. 1 (October 1969).

lesser evil, or perhaps even welcomed as one alternative counterweight. It is doubtful, however, that either the Japanese or the Southeast Asians would go so far as to allow Japanese ground forces to be sent to Southeast Asia for peacekeeping or other security purposes. It is more probable that the Southeast Asians would insist on obtaining Japanese military cooperation via regional channels such as the Association of Southeast Asian Nations (ASEAN). In any event, the initiative will remain in Southeast Asia.

The fourth dynamic factor is the emerging similarity of emphasis on the non-military requirements of Asian security. The idea is to shift from a primary reliance on military forces to what some of the Asian spokesmen call "political counterbalances" or "political counterpoises" to lessen tensions, improve security, and resolve international problems in Asia.

Political counterbalancing, an important new consideration in Asian international politics, emphasizes direct or indirect political consensus and organization of the Asian states, rather than the aggregation of military power and the confrontation of military alliances. The concept reflects the Asian goal of seeking social harmony by inclusion of all members via consensus. On a regional scale Japanese leaders suggest "a viable community of nations . . . aiming toward the attainment of a harmonious and stable whole."[14]

In a nebulous way, the Asian concept of counterpoise means using the solidarity of the weak to disarm and win over the strong in the sense of Zen or *aikido*. By pooling resources and demonstrating solidarity, Asians hope to create an entity to counterbalance the influence of China or of other large powers. Several interacting elements would build this counterpoise. The primary emphasis on mutual help for economic growth and social justice would decrease the dangers of instability, insurrection, and insurgency. The interdependence of

[14] Kiichi Aichi, "Japan's Legacy and Destiny of Change," *Foreign Affairs*, Vol. 48, No. 1 (October 1969), p. 34.

an indigenous community of Asian nations open to China would dilute alien, non-Asian content and discourage Chinese attack. Asian "political balancing" would provide a common forum for exchanges with China and might even persuade China to work with or join the group. In any event, an organized counterpoise with power, prestige, and unity would put the Asian states on more of an equal basis to deal with the great powers.

Thanat Khoman of Thailand calls this "a new concept of political security . . . founded on concerned and coordinated political actions . . . an Asian concert." He explains:

"We thus hope that first by organizing ourselves, by making our grouping an operational one, we shall offer a kind of attraction to the other side to come and work with us and to derive benefits from cooperation and neighbourly relations. Secondly, we have to go and meet the threat and the danger at the source. We cannot afford to hide ourselves in a fortress. We shall ourselves have to contribute whatever we can to resolve the problems and the difficulties or together with our friends who share the same idea. Therefore there is a possibility that, in the future, not only Thailand but others who find themselves in the situation of being threatened by certain regimes in this part of the world may go together to meet with the antagonists and talk things over and try to find ways and means to live together as peacefully as possible.

"Therefore we believe that there is only one solution. Namely, to close our ranks, to try to work out a new concept of Southeast Asian or perhaps Asian partnership; to form a new entity or new grouping that will be able to deal effectively more or less on an equal basis with Asian powers as well as with outside powers from Europe or America."[15]

Within this concept, a division of "military labor" could

[15] Thanat Khoman, "Post-Vietnam Period—A New Era for Asia," Press Release No. 23, Permanent Mission of Thailand to the United Nations, April 24, 1969, pp. 3-5.

indirectly provide a multiple balance of power for security. First, it is assumed that American power would be available and creditable to deter large-scale conventional aggression or nuclear threats, for, as Japan's Foreign Minister Aichi put it, "the regional military balance in East Asia is essentially that between the United States, our ally, and communist China."[16] Secondly, local forces from Japan to Australia independent of American support or commitments would assume responsibilities for their respective national defense. Thirdly, the emergence of a viable indigenous Asian community including Japan would provide mutual assistance to safeguard both the stability of frontiers and the integrity of member nations. That aid could be in the form of training, equipment, troops, or diplomatic support.

As to what form "political security" would take, Thanat and others suggest that formal treaties, elaborate organizations, and detailed commitments would probably not form the basis for regionalism. Instead, concrete interests, specific undertakings, and a shared concern would bind the participants together in a general way. Asian pragmatism rather than Western legalism would generate reciprocity. The new formula would thus tend to reduce military relations and expand commercial relations with the big powers and others outside Asia.

The proponents of new concepts are aware of the difficulties and dangers of "a diplomatic and political balancing of China." Peking, moreover, may even harden its hostility toward the countries of the south if they solidify their cooperation and especially if Japan becomes the effective power in Asia. The Chinese may disdain the Asian open door. Over the long run, political security may turn out to be neither a deterrence nor an attraction. If the military counterweight of another great power is forefeited or foreclosed, reliance on a purely Asian balance may backfire. On the other hand,

[16] Aichi, "Japan's Legacy," p. 36.

there is something to be said for a really indigenous, independent Asian organization influencing Chinese leaders in Peking not to challenge it militarily but to accept its reality in coexistence and non-aggression.

Four different "threats" could endanger Southeast Asia in the next few years: China, North Vietnam, intraregional hostilities, and internal revolution or breakdown. Regarding China, a difference in perception and assessment of security threats has existed in the past and may continue during 1970-1975. On the one hand, being relatively small (except for Indonesia), weak, and close to China, the countries of Southeast Asia have tended to view a strong and united China as a potential threat to their survival as independent aligned or non-aligned nations. Accordingly they have sought a counterweight to or accommodation with China to guarantee their protection and deter Chinese attack. The Japanese in 1970, on the other hand, have not viewed China as the major or likely security threat in conventional terms to Japan or Southeast Asia. To the Japanese the real threat there is increased insurgency and subversion supported by China and perhaps financed by some of the overseas Chinese in Southeast Asia. Meanwhile, many Southeast Asians are also coming to feel that the conventional threat of attack by land or invasion is receding and that the real danger is Chinese-supported insurgency.

The potential nuclear threat from China fortunately will not take place during the early 1970's. It is difficult as yet to see how the Southeast Asians would perceive the emergence of Japanese nuclear capabilities to checkmate the Chinese "blackmail" threat against such targets as cities in Southeast Asia or against Japan and the United States. It is too early to tell how the Southeast Asians evaluate that nuclear threat and whether they will favor Japan's becoming a power as an offset. If nuclear parity and stable relations between the two nuclear superpowers can continue to provide a kind of global power balance, the probability is that many Asian governments will for some time oppose Japan's going nuclear to meet a Chinese threat. First, there would be many great dangers in any such nuclear expansion. Second, the establishment of an Asian political union or forum to include China might be preferable to nuclear proliferation, at least at first.

Union with China would imply some mutuality of goals and interests, making China's nuclear superiority seem less relevant and even somewhat obscure in this Asian context. China's presence—as at the Bandung Conference of 1955—would provide for direct exchanges and even some accommodation of viewpoints. As outlined above in relation to China, the formula of "political security" would serve as kind of Asian quid pro quo, a balance of accommodation, to use the phrase of Indonesia's former ambassador to the United States or a convergence beyond containment. In return for China's participation in an "Asian concert," other Asian countries would oppose nuclear expansion. Japan could forego this option. They all could agree on converting Asia into a nuclear-free zone guaranteed by mutual non-aggression and non-intervention pacts, including the two superpowers. Otherwise, in the absence or failure of some such collective approach, some Asian countries will probably favor either a joint U.S.-Japan nuclear role in Asia or an independent Japanese nuclear capability if China brandishes its bombs over Asia and backs its revolutionaries abroad.

Except for the nuclear factor, many of the same considerations apply to the threat from North Vietnam. Neighbors of North Vietnam have long had, and still have, actual experience with Vietnamese expansionism over the centuries. Laos, Cambodia, and Thailand know what it means in reality. Malaysia, Singapore, and Indonesia, although geographically farther away, are also afraid of the extension of North Vietnam's authority and control over Indochina and northeast Thailand. On the other hand, Japanese opinion, being farther away and more immune, does not seem to attach so much importance to this North Vietnamese threat to the other countries of Southeast Asia. It is possible, therefore, that if the war ends in Hanoi's control over Indochina and Japan then strengthens its diplomatic and economic relations with the new regime, a certain political estrangement between Japan and the other countries of Southeast Asia may result. This would be especially true if Japan were to pursue this policy in spite of a North Vietnamese *cum* Chinese spillover into the rest of Southeast Asia. That Japan would do so, under these circumstances, seems extremely unlikely, although the possibility does remain that if the influence of Hanoi and Peking continues to rise in Southeast Asia as a whole, the Japanese may particularly not wish to challenge the Chinese.

It seems more likely, however, that even if Hanoi does take over the peninsula, the other countries of Southeast Asia will prefer to accommodate to the new facts of life, especially if the new arrangements are internationally ratified and widely accepted. They will be even more inclined to do so if the power of the new regime stops at Thailand's borders and does not threaten Malaysia, Singapore, and the Philippines but cultivates relations of "peaceful coexistence." In that event, an initially more accommodating and tolerant attitude toward Hanoi and its "confreres" would not remain a serious liability for Japan's economic and political relationships with Southeast Asia. Indeed the whole area, communist and non-communist nations alike, might come to look upon Japan as an acceptable "partner counterpoise" to China, the USSR, and the United States.

If, on the other hand, South Vietnam survives and increases its viability, it could decrease the communist threat to take over all Cambodia and Laos. That in turn would dilute the threat to Thailand and the rest of Southeast Asia and make it easier for Japan and non-communist governments to develop their interrelationships. At the same time such a patchwork in Indochina might perpetuate protracted hostilities for years and keep Southeast Asia indefinitely in a state of dangerous tensions.

Intraregional disputes or hostilities may seem a lesser threat of armed conflict, but one should not discount the possibility that territorial, ethnic, nationalistic, or political differences of a contemporary or historical character could still arise between two or more of the Southeast Asian states. The Philippine-Malaysian dispute over Sabah and Sukarno's confrontation with Malaysia were sad examples of this contingency. In both cases Japan tried to mediate behind the scenes; its efforts were not unwelcome in Southeast Asia. Consequently it is reasonable to suppose that Japan and the Southeast Asian countries would view the intraregional threat in a similar way and collaborate to meet it without arms.

The fourth threat—and perhaps the most real one—is political revolution or social breakdown. There is a general view in Southeast Asia and Japan that Southeast Asian countries are uncohesive, volatile, unstable to a lesser or greater degree. They may suffer from political revolutions, of one group's overthrowing another, or from social revolutions such as may occur in the Philippines if conditions of poverty and injustice become unbearable. Political revolution and social breakdown in Thailand and Indonesia, however unlikely they may seem at present, would severely hurt the security of Southeast Asia and inhibit the development of any security arrangements between Southeast Asian countries and Japan in a new Asian balance of power. Communist insurgency is not the only internal security threat. Minority groups—tribes, religious sects, linguistic or regional communities, or new urban have-nots—could easily arm themselves to harass or rebel against the central authorities and each other. But it is doubtful whether, should such revolution or chaos occur,

Japanese action would be practical. In any event, it is extremely unlikely that it would be offered, invited, or accepted.

Different perceptions of threats arise partly out of the different positions of Japan and the Southeast Asian countries in Asia and on the world scene. Unlike the southern rim of China, Japan does not seem threatened by conventional attack by air or sea. Nor do subversion and insurgency seem a likely threat to Japan's stability and economic resurgence during 1970-1975. Japan thus seems reasonably immune to any of the four threats we have been discussing except for the possible contingency of a nuclear threat from China in the late 1970's.

More to the point, Japan will be developing a much different world position from Southeast Asia. Japan is becoming a global economic state and one of the most powerful components of an increasingly internationalized economy. Japan is also becoming involved in major world and non-Asian issues of security, environmental control, and the world balance of power between the two superpowers and the other large countries. The Japanese will continue to maintain or develop economic, cultural, and political ties with Western Europe and the Western Hemisphere, particularly the United States. Thus Japan is a multi-dimensional power—Asian, Western (technologically), and global. None of the Southeast Asian countries—in fact no other Asian country, including China—will parallel Japan in these respects for years to come, and they may resent their inability to do so.

In sorting out these different dimensions of Japanese interests, Japan will give different weight to questions of security threats in Southeast Asia than will the Southeast Asians if the Japanese feel they must balance off their Asian, Western, and global stakes. On the one hand, Japan possibly will first consider itself an Asian power and place highest priority on assuring the stability of its hinterland and some economic interdependence with many or most of its Asian neighbors.

Playing the role of primary Asian power could be expected to strengthen Japan's position with the advanced countries particularly and throughout the world generally. On the other hand, Japan may come to believe that its non-Asian stakes far outweigh its Asian stakes. In that case Japan would decrease its involvement along all three lines in Southeast Asia.

The Southeast Asians, communist or non-communist, will express increasing concern and even alarm if Japan suddenly increases its military power and asserts its political role. The memories of World War II would quickly surface. Japanese discussion of increasing their conventional military strength has already intensified apprehension in Southeast Asia. Even when that discussion is limited to developing the defense of Japan itself, Southeast Asians fear that this military leadership will somehow spill out into other parts of Asia. So far, Southeast Asian leaders have resented and deplored the attitudes expressed by some American officials, journalists, and others that Japan should take over more or all of the responsibility for the security of East Asia and the balance of power in Asia as a whole. Combined with Japanese economic dynamism and increasing interest in taking political initiatives, the indications of a military buildup in the near future initially frighten the Asians. This is an issue not to be overlooked. It might even go so far as to unite Peking and its communist allies with non-communist governments in the rest of Asia if the Japanese were to rearm too abruptly and extensively. In other words, unilateral buildup of Japanese military power would not necessarily become a stabilizing influence or help to generate a workable balance of power in Asia. In fact, it could have just the opposite effect. Much depends on how the Japanese conduct themselves to relieve the fears and apprehensions of Asians and on how they treat Southeast Asia economically, politically, and militarily—particularly economically. Japan can move fast along the first, more rapidly along the second, and slowly along the third, but the interaction among these three relationships and between Japan and Southeast Asia will be delicate and complex at best.

So much may be said of the region as a whole. It should not be forgotten, however, that Southeast Asia is an area of great diversity. For example, Vietnam differs substantially from all its neighbors because its traditions and culture stem primarily from China. The Buddhist countries of Laos, Cambodia, Thailand, and Burma in many respects do not share the outlook and background of the Muslim countries and peoples of Malaysia and Indonesia, nor the Catholic orientation of the Filipinos. In political and security terms, Thailand, South Vietnam, and the Philippines are formally allied with the United States; Malaysia and Singapore with Australia, New Zealand, and Great Britain. Indonesia, Cambodia, and Burma are not aligned, while Burma abstains from almost all external participation. The forms of government vary all the way from the representative government of the Philippines, Malaysia, and Singapore, to a more unitary politics in Indonesia and Thailand, to the socialist forms of Burma and Cambodia, to the communist regime of North Vietnam. Economically the countries of Southeast Asia also vary considerably in population, per capita income, resources, and policies. Within this heterogeneous community, three nations stand out as bellwethers: Australia, Thailand, and Indonesia. Japan's relations with them are therefore of peculiar importance.

In the 1970's Australia will have a new and weightier impact on Asia than during the past twenty-five years. The reason is the shift that has taken place in Australian attitudes toward its Asian neighbors and in the state of Australia itself. Immediately following World War II Australia anchored its security on its alliance with the United States and a forward American military presence to contain communist China. As part of a "forward strategy" in Southeast Asia, Australia participated imaginatively and vigorously in collective security arrangements—ANZUS and SEATO—and in bilateral military actions such as took place in Thailand and Vietnam during the 1960's. Generally speaking in the initial postwar period Australia considered Southeast Asia—and most of Asia, in fact—a rather dubious, distant, and dangerous place.[17]

During this period Australians were concerned about a resurgent Japan as well. They opposed Japan's rearmament or its reestablishment in any major role in the Pacific or in Asia. These Australian attitudes toward the Asian countries in general and Japan in particular reinforced Australian reliance on the United States. In fact, Australia's strong attachment to ANZUS, the treaty with the United States, and Australia's vigorous and imaginative participation in the Southeast Asian Treaty Organization were originally undertaken in the early 1950's partly to obtain some assurance against any future threat from Japan as well. Indeed, the Australian government objected in 1951 when the United States proposed a comprehensive regional security alliance to include Japan and Australia as well as other countries.

But Australia has changed.[18] The continent is suddenly found to have unbelievable mineral wealth. Discoveries or estimates of sensationally huge mineral deposits and their rapid exploitation will soon turn Australia into one of the world's two largest exporters of minerals—bauxite, iron, lead, zinc, oil, and coal. Australian trade is shifting from old markets in Europe and Britain to newer markets in the Pacific,

[17] Bruce Grant, "Toward a New Balance in Asia: An Australian View," *Foreign Affairs*, Vol. 47, No. 4 (July 1969), p. 715.

[18] This synopsis of Australian attitudes is taken from T. B. Millar, "Australia: Changing Policies for a Changing Environment," *The World Today*, Vol. 25, No. 7 (July 1969), pp. 306-15; Gelber, pp. 223-28; Bruce Grant, pp. 711-20; and D. K. Palit, "Evolution of Australian Defence Policy," *India Quarterly*, Vol. 25, No. 3 (July-September 1969), pp. 216-28. The synopsis also draws heavily on official Australian statements contained in *Current Notes on International Affairs* issued by the Australian Department of External Affairs.

especially Japan. With both economies booming and benefitting from their relative security and proximity, economic complementarity is emerging.

In these changing times and circumstances, Australian leaders have described Australian policy concerning Asia in altered tones and terms. On February 25, 1969, Prime Minister Gorton set forth the basic regional policy of Australia toward "our neighbors of the north":

"Our starting point was and is that we are part of and are situated in the region. Hence security, stability, and progress for the other nations in the region must also contribute to the security of Australia. We cannot fail to be affected by what happens in our neighbours' countries. What affects their security affects our security. . . . We could not turn our backs on our neighbours, refuse to help provide forces for their safety, and wash our hands of the possible consequences."[19]

". . . we are prepared to maintain and are planning to maintain forces of all arms in that area after the British withdrawal—without setting any specific terminal date."[20]

In Prime Minister Gorton's conception, regional stability and security depended on three things in the region itself: avoidance of territorial and other disputes; economic progress reflected in rising standards of living for the "ordinary people"; and peaceful cooperation between countries in the region in many fields.

Australia's regional policy is carried out along two lines: economic assistance and military support. Concerning the first, Foreign Minister William McMahon said: "We must therefore work in all ways open to us to reinforce those factors making for stability, and give all possible support towards cooperation and mutual help in the region. We must recognize the critical importance of adequate living standards

[19] Millar, p. 310.

[20] John Gorton, "Defence in South-East Asia," *Current Notes on International Affairs*, Vol. 40, No. 2 (February 1969), p. 42.

to the independence and progress of the countries in the region, and must continue to give aid within the limits of our capacity towards the economic and social developments of these countries."[21]

The Australian defense minister, Malcolm Fraser, translated the general regional policy into terms of security.

"The region of South-East Asia and the surrounding Pacific and Indian Ocean waters comprise our environment: We are as well a part of the environment of the other nations in our region. . . .

"Of course we and other countries hope that by diplomacy and policies of aid we will reduce and ultimately eliminate threats to the region so that we may all devote our energies to improving the standard of life of our people. Military isolation on Australia's part would obstruct this objective. Military co-operation is designed to establish security so that the governments concerned can work for their own people without hindrance. . . . We believe there will be no permanent security for any of the small countries of the region until there is permanent security for all. This being the case, within our resources our military capability must be geared for deployment in the region of which Australia is a part when in our judgment we conclude that this is demanded by our concept of regional security as well as for the obvious purpose of meeting possible threats to Australian territory."[22]

Within this regional context, official Australian attitudes toward Japan's participation have shifted considerably in twenty years from the former suspicion of a resurgent Japan to a current acceptance of Japan's significant reemergence. Australia and Japan are now extensively involved together in developing a huge market of their own, which reflects the

[21] William McMahon, "International Affairs," *Current Notes on International Affairs*, Vol. 41, No. 3 (March 1970), p. 102.

[22] Malcolm Fraser, "Defence," *Current Notes on International Affairs*, Vol. 41, No. 3 (March 1970), p. 140.

drastic technological and economic changes underway in both countries. Japan is buying huge volumes of Australian raw materials—some 60 percent of their export. Australia is promoting Japanese investment in industry in Australia and buying Japanese products.

The former foreign minister, Gordon Freeth, spoke for many Australians when he expressed the opinion in September 1969 that before long Japan would assume greater political responsibilities in the region. He pointed out that in the Asia-Pacific Council (ASPAC) Japan had given "considerable leadership in developing consultations in a relaxed and informal atmosphere on the political and strategic problems common to his region as a whole."[23]

His successor, William McMahon, also speculated in a parliamentary statement that the 1970's would see Japan "taking a wider and more active role overseas . . . in time, Japan's influence must inevitably extend beyond the commercial and economic sphere."[24] As to the effect on Australia, he noted that Japan's economic growth, aid to the developing countries in East Asia, and trade with Australia had both strengthened regional stability and benefitted Australia. Moreover, the minister emphasized, Australia had developed "increasingly close contacts with Japan in the political field." Regular consultations are periodically held at a high level concerning disarmament, regional affairs, the United Nations, and China, among others. As to Japan's part in regional security, he indicated support for the prospect that Japan would not assume any military role beyond the defense of Japan itself, but could make "a decisive contribution to the security of the area by promoting industrial and commercial growth." Australia, he said, would welcome and do its best "to encourage her par-

[23] Gordon Freeth, "American and Australian Relations with Asia," *Current Notes on International Affairs*, Vol. 40, No. 9 (September 1969), p. 527.

[24] McMahon, "International Affairs," p. 97.

ticipation in the consultations that are becoming increasingly important in the political life of the region."[25]

Notwithstanding the clarity and coherence of these official statements, some ambivalence, divergence, and confusion have also emerged concerning Australia's foreign relations. A forward position in Asia and the Pacific is criticized while reliance on the United States is questioned in some quarters. The growing dialogue is reexamining Australia's role and Australia's treaty relations in the balance of power and the security of East Asia during the 1970's. As one Australian observer put it: "The question is rather whether military containment based on American power is more effective than a neutral independent Australia would be in controlling China. Once the policy has shifted from containment to a diplomatic and political balancing of China, the regional issues formed in the last few years should develop some muscle."[26]

Finding itself in this fast-moving "new era" which requires the search for new security policies and the retention of old security practices until they can be replaced, Australia will probably follow several lines of action in the 1970's. First, it will try to keep its alliance with the United States and American commitments in SEATO in good working order, hoping that the United States will not entirely withdraw from a forward position in Southeast Asia and that an Australian consensus will continue to support that American relationship. Second, Australia will adhere to its significant commitment to the defense of Malaysia and Singapore. This is essentially an arrangement within the Commonwealth of Nations and is not yet directly related to ASPAC or ASEAN. Yet it could provide an indigenous nucleus for moving toward a broadened new security system. Third, Australia will probably not try to persuade Japan to undertake any security or military commitments in the area until there is a much more reciprocal acceptance of such Japanese participation. Mean-

[25] *Ibid.*

[26] Grant, p. 719.

while, Australia will continue to develop close political and economic relations with Japan, encourage economic interdependence in the Pacific Basin, and support the Japanese-United States security alliance.

Thailand will be particularly significant in the security of Southeast Asia in the 1970's as a potential flashpoint for another large-scale insurgency, particularly if Hanoi wins control of Laos, Cambodia, and Vietnam. The war there—particularly in Laos—represents a danger for Thailand that may continue for years. By mid-1970 external insecurity surrounded Thailand along *all* its land frontiers—Burma in the west, Laos in the northeast, Cambodia in the east, and the Malaysian border in the south. Because of this geographical insecurity and because of Thailand's nearness to China and North Vietnam, Thailand faces the potential threat of foreign armed aggression across its frontier or externally supported indigenous insurgency and guerrilla-type harassment on a major scale.

From the standpoint of historical precedent, the Thai and the "Tonkinese" have often vied for mastery in Southeast Asia and generated shifts in the local balance of power. Now most observers are inclined to conclude that North Vietnam will not invade Thailand by conventional forces in the next few years, nor will China, which never did so in the past. But their direction and support of Thai-based insurgency could substantially increase. If North Vietnam wins military and political power in all of Indochina, Thailand's border areas along Laos and Cambodia will be easily infiltrated and perhaps even subverted. Hostile control of western Cambodia, so close and accessible to Bangkok, would endanger the heart of Thailand. It is well to remember that the Japanese army quickly invaded the kingdom from Cambodia on December 7, 1941. If conditions also worsen in northern and eastern Burma, Thailand could be under siege. Meanwhile, China may put a variety of pressures on the Thais to try to make them more friendly or amenable to China, disengage them from the United States, and change their government.

If Thailand is the flashpoint for the early 1970's in Southeast Asia, the next question is what security measures the Thais will wish to take to promote their independence. Assuming that the United States government does not entirely rescind its commitments to Thailand under SEATO during the next five years, the Thais will probably count on that treaty and American armed support in the event of conventional armed aggression or nuclear threat. As they have been, the Thais will continue to rely on their own military and police forces to deal with insurgency, depending on the United States and some other countries only for supplies, equipment, and training. Only if the scale of North Vietnamese or Chinese support for insurgency becomes so substantial and intolerable that the Thais cannot handle it alone will they invoke the Southeast Asia Treaty for unilateral American intervention or combat forces or a collective action by as many members of SEATO as would respond. Under this assumption, Thai military bases and depots for pre-stocked military supplies would remain ready for allied use in the event of either contingency.

Under these circumstances it seems unlikely that Thailand would ask Japan for air, naval, or ground forces—if they were available—to help Thailand. By the same token, it is equally improbable that Japan would soon get involved in military operations in or on behalf of Thailand, despite the likelihood that Thailand is becoming an increasingly important market for Japanese purchases, sales, and investment. It is possible, however, that sooner or later Bangkok might also look to Japan for military supplies, in addition to the United States, Australia, and other countries.

But what if the SEATO alliance with the United States weakens and the Thai-American relationship deteriorates? The Thais will look for other options. Securing the counterweight of one power to offset the danger to Thai survival has long been a cardinal and successful practice of Thai leaders. If the United States ceases to look like a dependable counterweight, the Thais will seek another. Thus, the significant possibility arises that Thailand may seek some understanding or even entente with Japan during the 1970's.

A Japanese-Thai entente does have some precedent. Although the circumstances were very different from today, the Thai leadership did have such a relationship with Imperial Japan from 1935 to 1945. Now Japan again will become the major power in East Asia, certainly in economic and technological terms. Perhaps Japan will express a major voice about developments in the region, as well as on international issues. Japan might provide some counterbalancing resources if it sees a vital interest in the stabilization of Southeast Asia and if the United States appears to be receding from the area. Notwithstanding differences in culture, geography, and history, the Thais and the Japanese, who share some common traits, including a pragmatic approach to international relations and a close calculation of national interests, could well develop working relationships for mutual security, broadly but inclusively defined.

The Thais are unlikely to seek such a bilateral commitment except in a multilateral context in order to increase their room for maneuver. Thus, they would probably also seek to expand trade and cultural relations with the Soviet Union and East Europe to offset possible pressures from China, North Vietnam, and what might become a communist-controlled Indochina. At the same time, the Thai government would continue to let it be known that it favored improving relations and opening talks with Peking and Hanoi in order to lessen pressures on Thailand and the region as a whole.

But the "multiplicity" option that the Thai government would probably work on most vigorously would be an effort to generate a regional counterpoise with Japan by developing

ASEAN, ASPAC, or a new indigenous grouping into regional "political security" for mutual safety including Japan, but without rigidly excluding China or setting up a "bloc." Thailand has the diplomatic talent and experience to serve as the regional catalyst and geographic keystone. Thailand's diplomacy already has made an impressive record in East Asia. Bangkok has concentrated much energy to build up its ties with all its neighbors and to identify itself with the region of Southeast Asia. Visible and important, Bangkok has become the regional headquarters for a lengthening list of governmental and private organizations. The Thai government has also succeeded in becoming a respected diplomatic broker for dealing with regional disputes. With such good credentials, Thailand's concepts for organizing regional security have made a great impact in the capitals of Southeast Asia; in fact, it is fair to say that Thailand and its foreign minister, Thanat Khoman, have become the formulators of region-building concepts and initiatives for over a decade. It is only if these efforts did not seem to be producing an effective counterweight to growing Chinese or Vietnamese pressures that some Thai leaders might eventually favor a bilateral understanding with Japan. At the moment, this possibility seems remote, and the possibility that Japan would respond favorably, even more so.

Indeed, it seems equally possible, should the present security arrangements weaken, that the Thais will judge that a military counterweight cannot be found. In that case the alternative might be to seek security through a multiple political consensus among the countries of East Asia. This would present Japan with a different challenge, but one that would seem to fit better with its own reluctance to become militarily engaged abroad.

Indonesia's stability and security are also crucial for the region and for Asia as a whole. Indonesia straddles the long sea and air lanes connecting the Indian and Pacific Oceans.

Indonesia possesses vast natural resources and the fifth largest population in the world. Obviously this is a tempting strategic prize. Indonesia, however, does not face an immediate direct threat of external armed aggression or major insurgency supplied on a large scale from outside. Only if communist China and North Vietnam were to gain political and military control of mainland Southeast Asia, including Singapore, or if a revolutionary pro-Peking movement were to take over the Philippines would Indonesia be in real danger from outside pressures.

Indonesia's problems are basically internal. Accordingly, it will probably continue to emphasize its international nonalignment and its domestic development. It is only if the north-south geopolitical axis of Southeast Asia is really menaced and if the United States fully disengages from Southeast Asia that Indonesia might turn to Japan for military assistance and even military commitments.

Indonesians are aware, of course, of the military potential of Japan. Ambassador Soedjatmoko has remarked: "It is conceivable . . . that depending on the security development in the Indian Ocean and along the Western Pacific after Vietnam, Japan may at one point feel the need to assume a more direct security responsibility. Once embarked on this course, however, the mere existence of China's nuclear capability will make it impossible for Japan not to go nuclear as well. This in turn will compel her to move out from under the American defense system and to adopt a political and defense posture of her own. The profound impact of such a development on the balance of forces in the region hardly needs elaboration."[27]

Barring these eventualities, Indonesia and Japan will both try to follow similar policies for seeking a peaceful equilibrium or balancing of power in Asia along the lines of parallel

[27] Ambassador Soedjatmoko, *Indonesian News and Views,* January 19, 1970, Washington, D.C., Embassy of Indonesia, p. 4.

regional approaches. As President Suharto indicated at the Djakarta conference of May 1970, the Indonesians also will be inclined to avoid an "anti-communist crusade" abroad and sharp confrontation with China such as would be necessary in an American-supported Asia containment policy or in joining an Asian coalition against China. The Japanese and Indonesians will probably try to develop both bilateral openings to China and regional organizations that will invite and perhaps even attract Chinese participation someday. Like other Asians, the Indonesians and Japanese will seek non-military means to restrain Chinese pressure on the rest of Asia and induce Chinese cooperation.

The Indonesians will be reluctant, however, to encourage or allow Japan to become the predominant spokesman or initiator of Asian security arrangements. Likewise, they will continue to seek Japanese trade and aid for the development of Indonesia. Foreign Minister Malik has even proposed a Marshall Plan for Asian countries with a large role for Japan.[28] The Indonesians and the Japanese will probably find much in common if developmental assistance can be worked out in a framework of mutual cooperation and "harmonious interdependence." The Indonesians do not want too heavy a Japanese hand nor too much Japanese investment primarily to benefit Japan.

Thus, throughout Southeast Asia two processes seem to be at work. One is the steadily increasing economic involvement of Japan, supported by a growing but still very cautious political concern. The other is the gradual convergence of Japanese and Southeast Asian views on a conception of regional security that deemphasizes the military. There is some apprehension of Chinese land and Soviet naval power, but the chief threats to security in the 1970's are expected to arise from domestic instabilities. To these, it is felt, the most realistic answers are not to be found in military alliances with great

[28] *Far Eastern Economic Review*, May 14, 1970, p. 16.

external powers—a relationship from which, in any event, the United States is withdrawing and which the Japanese are neither asked to assume nor want to. Japanese and Southeast Asians agree that the best answers are to be found rather in a strengthening of the political, social, and economic fabric of each country, and that this can best be strengthened internationally by economic policies of aid and trade, and diplomatic policies of accommodation and consensus. These are the overseas security policies most countries in the region want Japan to pursue and, certainly through the early 1970's, these are the policies toward which Japan's economic involvement is leading it.

CHAPTER VII

A Forecast with Recommendations

JAMES WILLIAM MORLEY

THE major influences that will shape Japanese security policy over the next few years are now clear: a shifting international environment, increasingly confronting Japan with *realpolitik*; a rising spirit of new nationalism at home, calling restlessly for imaginative initiatives; and a continuing high rate of economic growth, which makes real alternative choices possible. How will Japan respond? The answer will depend in large part on the perceptions of Japan's conservative leaders, for it is they who will be making the decisions at least through the middle of the decade.

Deeply imbued with the doctrines of "economism" and "balanced defense," they are inclined to think of Japan's security in terms of three concentric zones: the home islands, the western Pacific, and Southeast Asia and beyond. In the zone of the home islands, for years the principal threat has been perceived to be subversion, anticipated to be a take-over of the government by communist elements instigated and supported by the Soviet Union and the People's Republic of China. In view of the accelerating failure of violent tactics, from the "Food May Day" in 1947, through the anti-treaty demonstrations in 1952 and 1960, to the student disturbances of 1968-1969 and the final fizzling out of the anti-treaty campaign in the spring of 1970, the possibilities of a violent take-over are increasingly remote. The possibility of a peaceful take-over by the ballot is slightly higher now that the Japan Communist party has adopted peaceful, parliamentary tactics, but certainly there would appear to be no realistic chance that a communist government or a coalition government with communist participation could come to power in Japan within the next five years.

Conservative leadership, however, is now and can be expected to continue to be, extremely sensitive to the revolutionary potential within Japan and to take this into account when shaping its foreign and security policy.

The possibility of direct external military attack on the home islands is judged to be much less of a danger. Only two countries are currently capable of launching such an attack: the United States and the USSR. Having decided more than twenty years ago to solve this problem by aligning with the United States, Japanese conservatives evaluate the threats from these two quarters very differently.

Historically, the image of the Russians is of a people threatening Japan's rightful place in Northeast Asia. Today this image is supported by the memory of what is felt to be Soviet violation of the Neutrality Treaty of 1940; Soviet acquisition of the northern territories after World War II, including the Shikotan and Habomai Islands as well as the "Southern Kuriles," which nearly all Japanese hold to be indigenous Japanese territory; the systematic squeezing out of the Japanese fishing industry from the northern waters in which it has traditionally operated; the aerial and naval surveillance which the Soviets attempt to maintain in the vicinity of Japan; and of course support for what is felt to be communist subversion of Japanese society. The long-range political objective of the Soviet Union is assumed to be to neutralize Japan, wean it away from the American alliance, communize it, and draw it into the Soviet "camp."

The fact that the Soviet Union has not taken more aggressive steps toward Japan in the postwar period the Japanese conservatives attribute primarily to the "balance of terror" established by the United States and the total balance of power in the world between the two "Cold War camps." So long as the United States gives convincing evidence that it retains the capacity and the will to retaliate against the Soviet homeland should the USSR threaten Japan overtly, Japanese conservatives do not believe the USSR will attack. It will, in fact, seek to dampen or attenuate conflicts and pursue a policy of peaceful coexistence.

The security problem vis-à-vis the People's Republic of China is seen somewhat differently. The conservatives acknowledge that the PRC, like the USSR, would like to see Japan communized and aligned with *it*; but they have never shared the view that the communist leadership in China—or any other leadership there, for that matter—would be able very quickly to transform that enormously complex and underdeveloped society into a modern state. They have felt some guilt for the China war and are convinced that they could provide many of the goods and services the Chinese will need to modernize effectively, but they do not feel that they were defeated in the Pacific war by the Chinese, and, particularly in view of the weakness of Chinese naval forces, they do not see any danger of a sustained Chinese attack on the Japanese home islands now or in the next five years. The possibility of a Chinese nuclear strike by low-flying bomber or suitcase has been recognized to exist for some time. The nuclear threat has obviously increased with the recently announced Chinese deployment of IRBMs. But the Japanese conservative leaders do not believe that Chinese military doctrine calls for assaults of this kind overseas; and in any event, particularly until the Chinese deploy ICBMs and thus threaten the American homeland directly, they feel confident that the American nuclear deterrent is sufficient, in this case as in the Soviet case, to restrain the Chinese from such adventurism. In their eyes, therefore, the most significant threat from the Chinese is not an overt attack but subversion.

As the decision to align with the United States shows, the conservatives, and indeed most Japanese, see many pluses in the American relationship, but it is also true that to varying degrees minuses also are seen. They are conscious of a sense of dependence and therefore vulnerability toward America.

This stems from the fact that the United States is Japan's largest overseas market, its chief source of foreign capital, its primary source of foreign technology, its principal source of nuclear deterrence to the PRC and the USSR, and also a principal supporter of the world order in which Japan is prospering. This sense has led them to feel they must conciliate American opinion in security as in many other fields more than they should like and therefore to exercise less autonomy over their own future than they desire. They have some concern lest the American ties may somehow inhibit the development of their own independent strength, keeping them permanently in a "younger brother" status. In addition, they have always had a certain measure of fear that the concession of base rights to the United States may draw them into a war more for American than for Japanese interests. And they have worried considerably about the strength that the opposition forces in general and the revolutionary forces in particular have derived from the unpopularity of the American military presence and the widespread fear of nuclear war, which the United States tie symbolizes for many.

These perceptions that neither the Soviet Union nor mainland China poses a threat of overt attack in the near future are reinforced by the Sino-Soviet conflict. So long as the Chinese and the Russians are heavily engaged with each other—and the Japanese expect them to be for some time to come—neither one is thought to want to divert its military attention to Japan. As Yasuhiro Nakasone remarked when director general of the Defense Agency, for Japan the Sino-Soviet conflict is "a welcome gift."

Japanese conservatives are also increasingly concerned with the security of the wider environment in the western Pacific, a region that might be defined roughly as stretching from Guam westward to the southern tip of Taiwan, thence northward to the thirty-eighth parallel in Korea, on to the northern tip of Hokkaido, and thence southward back to

Guam. Here the threats to their interests are seen to be chiefly three.

One is the threat to Japanese shipping posed by the increased activity of Soviet naval forces in the region. No overt act is anticipated in the near future, but the potentiality is beginning to give concern for the distant future.

A second is the threat to the Republic of Korea (ROK) posed by the Democratic People's Republic of Korea (DPRK) in the north and its Chinese and Soviet allies. As indicated by Premier Sato in his joint communique with President Nixon in the fall of 1969, the Japanese conservatives recognize that Japan has a vital interest in the ROK. While the nature of this interest was not spelled out, it may be presumed to consist of the economic stake that the Japanese have acquired and that can only be expected to grow. Some Japanese see an interest also in Korea's security. They seek, first of all, the preservation of South Korea at least in friendly hands and, secondly, in the preservation of peace in the peninsula lest the outbreak of war lead to engagement between the superpowers, thereby jeopardizing the balance of power on which Japanese security is thought ultimately to rest.

A third threat in the western Pacific is to the Republic of China (ROC) on Taiwan. In the same joint communique the premier indicated the concern of Japan for peace in this area. Presumably the Japanese concern is for the same stakes as in Korea: the preservation of the trade and investment that Japan has built up in the island, the preservation of the island in friendly hands, and the prevention of a war over Taiwan in which the superpowers would threaten the balance.

An element that makes the Japanese conservatives particularly sensitive to the security situation in Korea and Taiwan is the Sino-American relationship. Feeling little direct threat from the mainland and anxious in the long run to bring about a better inter-change with it, the Japanese government is particularly worried about the possibility that a resumption of war between the two Chinese regimes or the two Korean regimes would tend to drag in the mainland Chinese and the Americans on opposing sides, thereby drawing in Japan as well because of its responsibilities under the security treaty. On the other hand, President Nixon's visit to Peking in February 1972 has raised doubts in Japanese minds about America's long-run willingness to defend Taiwan. To the extent that this doubt grows, conservative Japanese leaders can be expected to become more ambiguous about their own concern for the Republic of China.

In Southeast Asia and beyond, the Japanese conservatives are very much interested in preserving an environment friendly to their trade and investment. In keeping with their perception of Soviet and Chinese military doctrine and the effectiveness of the American nuclear deterrent, they do not believe that the primary threat to the development and preservation of such a friendly environment—commonly called "stability"—is that of overt attack by the PRC or the USSR. Rather, the danger comes from the possibility of revolution and civil war within countries in the region and of border conflicts among them, resulting from their very recent emergence into statehood, the comparatively low level of their political and economic development, and the relative weakness in their cultures of certain social and psychological elements believed to be essential to sustained modernization.

This is not to say that the Japanese would be indifferent to a rapid pull-out of American forces in the Indochinese peninsula under conditions interpreted as an American defeat. The government of Japan has, in fact, consistently and publicly supported the American effort there; but, its support stems primarily from the desire for a continued American presence in Southeast Asia and concern for the reliability of American commitments, on which Japan so much depends, rather than from a belief that the conflict in the peninsula was instigated primarily by the Chinese or the Russians, that intervention by the United States was the best response to instability there, or that the security of Japan was in any direct sense threatened.

As for India, it is seen essentially as another, albeit the largest, of the underdeveloped countries to the south. Trade and investment opportunities are of interest, but the possibility of India's being able to modernize quickly and effectively is not evaluated highly and the security of India is not seen to have any direct relationship to Japan. Japan's most vital security concern in the Indian area is with the sea lanes to the Middle East, since Japan is absolutely dependent on them for its oil supply; but there is no expectation that India could or would protect them. In fact, India's acceptance of the Soviet Union's naval activity in that area makes them more vulnerable than before.

On the basis of these perceptions, Japan has formulated a basic security policy emphasizing four elements. The first element is the balanced development of Japan's total power stressing economic growth. In this conception economic growth has been seen as the necessary precondition for satisfying all of Japan's needs. Moreover, decisions as to how much any one need should be satisfied, whether it be defense, education, welfare, or governmental administration, have been taken with three criteria in mind: the relative strength of this need, the pace of Japan's overall economic growth, and the extent to which Japan's overall economic growth is promoted or at least not adversely affected by such allocations. The result has been a dramatic secular rise in Japan's GNP until it is now third largest in the world.

Fears have been expressed that this growth cannot continue. Some persons estimate that private consumption expenditure will take a proportionately larger share of the GNP as a result of an acceleration in the rise of the wage rate. Others believe that public consumption expenditures also will take a proportionately larger share in response to rising demands for education, welfare, anti-pollution, and other allocations. Others are concerned that the economy will be slowed by a decline in the labor force due to lower birthrates in the 1950's, the tendency for a larger proportion of the young to remain in school, and the gradual reduction in working hours. Still others estimate that the boom in private equipment investment so characteristic of the past twenty years may decline as the technology gap with other advanced countries is reduced. And, most recently, deep concern has been expressed that the new economic policy announced by President Nixon in August 1971 will result in a serious reduction of Japanese exports.

But the capital-output ratio is so high that several percent of the GNP could be shifted from investment to consumption, either private or public, without substantially affecting the growth rate. There are other offsetting possibilities: the labor force might be augmented by bringing into it more women and more elderly workers, both of whom are underemployed; the Japanese may well decide to invest more heavily in technological research and development; and the depressing effects of a possible decline in trade may be alleviated by increasing government investment in social overhead.[1] It is too early to know how effectively Japan will be able to adjust to the disorientation of the international monetary and trading system; but there is general agreement among qualified observers that unless there is a fundamental shift in policy orientation or in the international environment, the government of Japan should be able to sustain the high growth rate of the past for a number of years more, resulting in a doubling of the GNP between 1970 and 1975 to $470 billion and a quadrupling possibly by 1980.

[1] Announced by Finance Minister Mikio Mizuta on August 29, 1971, following news of the new Nixon economic policy. *New York Times,* August 31, 1971, p. 1.

The second element in Japan's security posture is a systematic build-up of Japan's military forces in keeping with the pace of economic growth. Defense expenditures have, in fact, been relatively modest, amounting consistently over a long period of time to less than 1 percent of the GNP and less than 10 percent of the national budget. In the government budget proposed for 1971 or in the Fourth Defense Build-up Plan (1972-1976) now being discussed, there is no indication of any inclination by the government to change the size of this carefully restricted share over the next five years. This does not mean that Japanese defense allocations will be inconsequential. Increasing as they can be expected to do at the same rate as the GNP, they will presumably go up about 16 percent or so every year (in current prices), doubling by 1975 to about $4.2 billion. Such expenditures would still perhaps be only half as large as mainland China's, and only half as large as those of Britain, France, or Germany, but they would be the second highest in Asia and presumably the seventh highest in the world.

Such a build-up should enable Japan to play a much greater role in its own defense and to exercise a greater degree of surveillance over its own and contiguous air space and over the surrounding waters, particularly between the home islands, the Bonins, and Okinawa and to a certain extent possibly as far as Guam and Taiwan. Its defense industry also can be expected to expand, so that by 1975 Japan will probably have stockpiled sufficiently for a reasonably prolonged war. At the same time, it will probably be engaging in a much-expanded program of non-lethal arms sales throughout the world, either on a strictly commercial basis or possibly with the support of government aid.

Government support for science and technology, including particularly those areas with defense implications—electronics, aircraft, space, and nuclear energy—can be expected to get more attention than in the past; but, for technological, budgetary, and policy reasons, by 1975 Japan is most likely still to be dependent on the United States for advanced weaponry and long-range protection, including nuclear.

Its forces will be small, well-trained, scientifically oriented, but limited in deployment to the home islands and surrounding water and air space. They will be a threat to no other country nor will they be capable of or available for assisting any other country, but they should be a substantial deterrent to attack on the Japanese islands themselves.

This modest build-up program is clearly based on two convictions: that, as explained above, Japan is under no immediate military threat; secondly, that time is on Japan's side, that is, the longer Japan refrains from a massive rearmament, the greater its capacity to do so on a crash basis if the need grows, and therefore, provided there is adequate warning for it to prepare, the greater in fact will be its military strength when and if a crisis does threaten.

In order to try to cope with the various levels of military threat that it does see and that it recognizes are beyond its own capacity to handle directly, especially during the next five to ten years, Japan places its reliance on two other elements in its defense posture: the American alignment and its "peace diplomacy." Under the mutual security treaty of 1960, which is a revision of that of 1951, Japan has accorded to the United States the use of certain bases to be used for the defense of Japan and to maintain security in the "Far East," meaning the Republic of Korea, the Republic of China, and the adjacent air space and waters. Because it would like to defuse the bitter issue that these bases have raised in Japanese politics and because it would like to acquire these same bases for the use of its own growing Self-Defense Forces, Japan has consistently sought their reduction, timed to the extent possible so that as the American forces move out, the Japanese forces move in. It would appear to be their hope that by 1975 or so most if not all of the remaining bases would be freed for Japanese use.

Thus, while keeping their own defense build-up at a graduated pace, they hope by their base policy to gain greater autonomy; but they do not wish to "go nuclear" or to take over any defense commitments outside of a restricted zone close to home. Therefore, it is important for them to secure the American withdrawal under terms and in an atmosphere that in no way weakens the American commitment to defend Japan with nuclear weapons should Japan be threatened with nuclear attack. The American nuclear umbrella, particularly against the USSR and to a lesser but growing extent against the PRC, is seen as absolutely vital.

Japan is anxious also that the American withdrawal from forward positions in Japan not be accompanied by an overall withdrawal of United States interest and capacity to act in Asia generally. The Japanese place great value on the contribution of American naval power to guaranteeing freedom of the seas for their commerce. They welcome the contribution of American forces to helping to maintain South Korea, Taiwan, and other areas in Asia in friendly hands. And they believe that American power generally is vital in Asia to maintain some kind of military balance with the Soviet Union and mainland China, on which the ultimate peace of Japan too is seen to rest.

This is not to say that Japanese defense policy is to encourage the United States to remain in a state of high confrontation in Asia or to enter into war there if it can at all be avoided. The government of Japan supported the United States in the Korean war. In the Sato-Nixon communique of 1969 it may be understood to have pledged to do so again should there be an overt attack from the north and should Japanese bases and presumably industrial capacity be required. It has also given public support to American efforts in the war in Indochina. But one must recognize that Japan's interest in American activity in Asia is essentially not in the United States' rolling back the countries with which Japan has or foresees some difficulties; in fact, it fears that any military engagements with such countries by the United States may in the end draw in the PRC or the USSR and escalate into the total war that it very much wishes to avoid. Of course, if the United States must fight wars in the Asia-Pacific area to maintain the balance in Asia or to protect its and Japan's friends, then the Japanese conservatives want it to win; but their greatest hope is in American deterrent power, that is, in the capacity of the United States not to fight wars but to deter them. This means that while Japan's conservative leaders would like to encourage the United States to remain in a military posture of readiness in Asia, they are also concerned that it not be provocative. On the contrary, they would like to see military readiness accompanied by a policy of diplomatic détente.

This would bring the United States more in accord with the fourth element in their own defense posture, Japan's peace diplomacy, that is, diplomacy designed to ease the confrontations and strains that cause wars. They are certainly not prepared to concede what they consider to be their vital interests in order to appease the communist states in Asia, and for the sake of the overall advantages of the American alignment they have not felt free to stray too far from the American stance; but within these limits they have tried not to antagonize anyone. They have, in fact, reached a general accommodation with the Soviet Union. They are interested in resuming relations with Outer Mongolia. With North Korea, mainland China, and North Vietnam they have permitted trade and, to varying degrees, cultural relations. They will want to go as far as they safely can with this détente policy in the years ahead, certainly, if it is at all possible, as far as the United States goes.

Peace diplomacy has come to mean also economic and technological aid, particularly to the Republic of Korea and to the other states of South and Southeast Asia, the objective being in part to support an expansion of Japanese trade and investment in this region, but in part also to strengthen these states so that they will be less prey to internal disorder or external attack. It is clear from numerous statements by government spokesmen that the Japanese conservative leaders consider such aid to be an effective alternative to military means for contributing to the security of the Asian-Pacific area. As Premier Sato told James Reston on August 31, 1971, "The money [Japan] would otherwise spend on bigger military budgets will go for economic aid to underdeveloped nations."[2] Symbolic of this conception is the suggestion that the level of aid and the level of military expenditures should be kept on a par with each other; and, in fact, this is clearly the policy to 1975, each being expected to reach about 1 percent of GNP.

While Japan is dependent on the external world for nearly all of its resources, the resource that is most critical of all—energy—is its real Achilles heel. Were imports of coal and particularly oil cut off, Japanese industry would grind to a halt within a few weeks. The conservatives have, therefore, considered it a matter of great priority to work to diversify their energy sources as much as possible. The long-run answer, they believe, is nuclear power and they feel that they can realistically hope to shift almost exclusively to it by the end of the century. But Japan has a long way to go. According to the Japan Atomic Energy Commission's 1970 White Paper, only four nuclear power generators will be in operation by October 1970; five more are under construction. In addition, the first experimental ship, *Mutsu*, should soon be equipped with a nuclear reactor. Moreover, these efforts are heavily dependent on the outside world, particularly the

[2] *New York Times*, September 2, 1971, p. 23.

United States. American corporations, such as Westinghouse and General Electric, are heavy investors, both of capital and of technology. And the United States, under an agreement signed in 1968 and continuing for thirty years, is the principal supplier of the nuclear fuel required.

This relationship enables the conservatives to move ahead with their plan but does not give them the independent capacity they are seeking. Consequently, a vigorous search is on for new deposits in various parts of the world that the Japanese can themselves exploit. In addition, with the establishment of the Uranium Enrichment Research and Study Group by the Japan AEC in May 1969, a program has been inaugurated to attempt to develop an indigenous enrichment capacity. The schedule calls for investigation of various processes, to be completed and a decision made as to method by 1972, and thereafter for construction of an experimental plant.

It is primarily because of what is felt to be the critical need of Japan for the most rapid and effective development of nuclear capacity for peaceful industrial uses that the conservatives have signed the Nuclear Non-Proliferation Treaty (NPT) so reluctantly and have delayed submitting it to the Diet for ratification. They fear that somehow in remaining dependent on the nuclear powers for explosive devices, they may limit their technology, and that in permitting international inspection, they may lose the commercial secrets of their research. They are not and will not be prepared to bind themselves in any way that might limit their ability to develop or apply nuclear power as they see fit.

They are not unmindful also that complete dependence on the United States for nuclear security limits their international policy options. Their delay in ratifying the NPT stems in part also from their reluctance to perpetuate this dependency. On the other hand, the present leadership does not want Japan to manufacture nuclear weapons, acquire nuclear weapons, or permit the United States to bring them onto

Japanese soil once Okinawa is returned. They believe that, largely due to the extraordinary concentration of Japan's population and its geographic proximity to Soviet and now Chinese nuclear delivery systems, a modest second strike capacity would have little meaning. Japan is unusually vulnerable to nuclear attack. To be able to deter such an attack by threatening to cause comparable damage to a potential enemy with a highly dispersed population and with a less concentrated or sophisticated industrial and communications establishment, such as the PRC or the USSR, would seem to require an overwhelming second-strike capacity on the order of that possessed by the United States or the USSR. That for the foreseeable future would be too expensive for Japan. Such a nuclear program also would so confront public sensibilities as to divide Japanese society further and jeopardize the continuance of the conservatives in office. It would also shock and antagonize the rest of the world, undermining Japan's peace diplomacy. They are also aware that unless and until Japan acquires a relatively secure, independent source of nuclear fuel and its own capacity for enrichment, its nuclear dependence on the United States would continue even if it began to make its own weapons. Finally, they are aware that a nuclear weapon is only as effective as its delivery system. Japan remains dependent on external suppliers, principally the United States, for its jet engines and for advanced electronic technology. It has a modest aircraft development and space program but at the present rate of progress is unlikely to have acquired a really independent aircraft capacity by 1975; and, having concentrated on propulsion rather than guidance systems in its space program, it is unlikely to have become independent in missiles by that date.

The odds are, therefore, that so long as they find the United States reliable, and at least to 1975, they will prefer to depend on it for nuclear protection, concentrating their own nuclear energies on peaceful uses, which in any event could be readily converted to war purposes if an emergency required.

It must be understood, however, that this inclination depends on the activity of the nuclear powers. From the Japanese conservatives' point of view, not only must the American nuclear umbrella remain reliable, but the gap between the "nuclear have" and the "nuclear have-not" nations must be narrowed. That is, for it to be tolerable in the long run for Japan to be without its own nuclear weapons, the importance of these weapons in world affairs must be reduced. The nuclear powers must show steady progress toward agreements to limit the further sophistication of their devices and to limit and soon to reduce their stockpiles.

The perceptions, policies, and inclinations of Japan's conservative leaders outlined above may perhaps best be characterized as "economism," by which one means to suggest the priority of economic growth in the national objectives. Among the allocations for other objectives, such as welfare, education, security, and the like, there is a sense of balance, a sense that carries over into the conception of security as resting on non-military as much as on military strengths. This stance was adopted by the political, bureaucratic, and business leadership that rallied around Premier Yoshida in the Occupation period and that has continued to dominate the thinking of the ruling segments of the Liberal Democratic party and its allies ever since.

While economism is likely to prevail to 1975 and has at least a fifty-fifty chance to continue thereafter, one must recognize that the party that champions it no longer commands a majority of the popular vote. It has, in fact, been slipping in popular esteem ever since its founding in 1955. Even if the decline continues at the present rate, the party will probably be able to so organize its voter support as to continue to secure a majority of Diet seats through at least the next two elections, presumably through 1975; but its success in this, and particularly thereafter, depends on its ability, which has not yet been clearly demonstrated, to build a new base of support in the cities. To do this presumably it will have to perfect a new technique of organization other than the alliance of local notables and the appeal to traditional ties that proved so effective in the countryside. This in turn may well require it to respond more than it has known how to do or than it has been willing to do to two broad sentiments on the rise. One is the rising sense of frustration of its city dwellers, who are plagued like urbanites elsewhere with a deep sense of isolation and powerlessness and a deep resentment at the inadequate housing, overcrowded transportation, and polluted environments that they must endure. The other is the rising sense of frustration on the part of nearly all Japanese that somehow Japan does not have the "independence" and the international prestige that it deserves. This is not to say that the Japanese people do not take great pride in their rising GNP. They do; and it will be politically essential for the conservative leaders to maintain it at something like its present growth rate. But they will need also to respond more effectively to these other pressures if they are to be able to remain in office.

The continuance in office of economism's supporters would also appear to depend in part on the international environment. Their most fundamental appeals are, after all, that economism is the best path to two very popular goals: peace and prosperity. Any international development that jostles the four-power balance between Japan, mainland China, the Soviet Union, and the United States, upon which Japan's security is now seen to depend, will have a direct effect on the fate of economism in Japan. The outbreak of hostilities, such as the renewal of the war in Korea or a degeneration of either the Sino-Soviet or Sino-American confrontation into war, would be likely to plunge Japan into a deeply divisive public controversy in which economism and its champions would lose out. A worsening of American-Soviet relations would have the same effect as would the worsening of Japan's relations with any one of these three powers. This latter possibility includes economic as well as military relations, and specifically the possibility of a down-turn in the world economy or of a rash of restrictive trade and investment policies being adopted by the United States or other major trading partners abroad, so that Japan's economic growth is severely restricted. The same negative effects could also be expected from any drastic change in the military capacity of any one of the powers vis-à-vis the others, as, for example, would be produced by a rapid deployment of IRBMs or, more surely, ICBMs by the Chinese, a naked expansion of Soviet naval power in the western Pacific, or a widespread and rapid withdrawal of American concern for Asian, particularly Japanese, security and the capacity to give that concern weight.

On the other hand, an easing or improvement of relations among the four powers—for example, between mainland China and the Soviet Union or between mainland China and the United States—may be expected to have alternative effects, depending on the circumstances. If a Sino-Soviet rapprochement takes place in a spirit of a general relaxation of tensions in Asia, including those between the two communist powers and Japan, the Japanese can be expected to welcome it, and economism—or whatever other policy is then in effect—will receive the credit. Similarly, if a Sino-American rapprochement takes place in full consultation with the government of Japan and at the same pace and in the same direction as a similar development in Sino-Japanese relations, that too will have a very positive effect. But if, on the contrary, a Sino-Soviet rapprochement takes place in a spirit of rebuilding the bloc to oppose Japan, then of course the result will be negative. Likewise, if the United States continues to negotiate secretly with the Chinese or proceeds at a pace or in a direction that the Japanese are not prepared themselves or permitted by the Chinese to follow, then too the economist policy will be severely attacked.

Were the economists unable to cope with the domestic pressures challenging them or the international threats that could arise, it is still possible that, since issues are not crucial in the factional struggle for power within the party, they would be able to retain their hold on the government but would simply shift their views. It is also possible, and indeed likely, that this would not work. One would suppose that, depending on the depth of the party crisis, the first step would be for a new factional alignment within the LDP to take over, one that would orient itself around somewhat different policies. The second, more serious step if the party seemed to have lost its future would be for various conservative factions to break off and join with all or some of the factions now in opposition under the Democratic Socialist, Komeito, or Socialist party rubrics to form a new governing center with a decisively different stance.

One cannot predict safely what alternative policy configurations would eventually be adopted, but certain possibilities can be identified.

First, disarmed socialism no longer seems realistic. For nearly twenty years the Socialist party has looked on itself and has been judged by observers to be the chief opposition in Japan, and the public debate has often revolved around the ideas championed by the JSP and those of the LDP. Domestically the JSP has stood for "socialism." Internationally it has stood for disarmed neutralism. But these ideas have increasingly lost their appeal. Doctrinaire socialism has hardly seemed able to promise more jobs or more rapid growth. The failure of the Great Leap and the Cultural Revolution have dimmed the confidence so many Japanese formerly felt they could place in the mainland Chinese. Japanese self-confidence is rapidly returning and the maintenance of the Self-Defense

Forces, at least at their present level, has come to be widely accepted. These facts plus the aging of the JSP leadership and its inability to organize the floating vote of the cities help to explain the decline of the Socialist party and the transformation of disarmed socialism into one of the might-have-beens of history.

While doctrinaire socialism, neutralism, and disarmament no longer represent an attractive pole around which a viable opposition can rally, a new cause is attracting support. It may be called broadly "welfarism," the central proposition of which is that the national priorities should be shifted. Economic growth is fine, but it is not enough. It does not seem automatically to lead to the better life that the Japanese people have been striving for so sacrificially for the past one hundred years. If Japan is really doing so much better, then it is time to share the wealth a little more—time to improve the social welfare system; put higher education on a stable financial basis; build the houses, the roads, the sewers, the public transportation systems that are needed; clean up the environment; and enhance the cultural life so that all those who have borne so heavily the strains induced by Japan's phenomenal growth—especially the farmers, the small shopkeepers, the commuters, the urban migrants, the youth, and the older generation—will share more fully in the better life that should be possible. In short, the national priority ought to be shifted from the quantity of Japan's income to the quality of its life.

There is no reason why a skillful government cannot reallocate several percent more of the GNP to these worthwhile causes than it is doing without really hurting the growth rate. Indeed, it would seem highly advisable for it to do so. But that is not to say that it will or, if it does, that it will be able to devise programs that effectively relieve these frustrations. If it fails, the political pendulum may swing even further, with the result that new leaders may take over in a coalition government of the center-left. If this happens, one may anticipate a configuration of attitudes affecting security policy along the following lines. The growth rate will probably be allowed to decline. Allocations for welfare in its broadest sense will go up drastically, since this sector will have priority; those for defense will probably be cut, in share of national budget if not in absolute amount. At the same time, with a slowing economy, foreign trade will decline and foreign exchange reserves will grow at a much slower pace. Accordingly, Japan will be able to exert its influence abroad less effectively than under economism and indeed will be less interested in doing so. It will curtail its concern for good and close relations with the United States and will probably want to terminate the security treaty, the space cooperation, and other arrangements, and to restrict American business, technological, and cultural relations in Japan. It will also have less concern for the development and security of non-communist Asian states, probably reducing its aid, tariff preferences, and other forms of assistance. To the extent that it does look abroad, it will concentrate instead on achieving a rapprochement with mainland China and other communist countries.

A third alternative might be designated "chauvinism." Nationalistic sentiments are obviously stirring in Japan. The economists are responding to them in a restrained way, and there is no clear reason why they cannot accommodate them in the economist stance, but again there is no guarantee that they will succeed. The public mood or international circumstances could shift abruptly. Were this to happen and were a leadership to come to power insisting that not economic growth and not welfare, but arms and prestige, should have first claim on Japan's national efforts, one may anticipate a configuration of attitudes along the following lines. The share of the budget allocated for defense purposes would be doubled or tripled. An effort would be made to crush the opposition to a revision of the constitutional restraints in Article 9. An effort to introduce conscription would be made. Large-scale science would be heavily funded in an effort to become autonomous in aircraft engines and missile guidance systems as soon as possible. The security treaty with the United States, whether abrogated or not, would have lost much of its meaning. Nuclear weapons would be added to the arsenal. The NPT and other strategic arms control measures would lose their appeal. By 1975 or before Japan would presumably have an armed force on the order of that of France or mainland China.

Such changes could come about only in an atmosphere of popular patriotism or government suppression as to give much greater currency to emotional chauvinism. This would heighten the tension in Japanese society, since certain freedoms inevitably would be curtailed and certain demands for relief from the ills of urban, industrialized living would have to be ignored. In addition to greater instability at home, such a shift would cause great consternation abroad. The communist powers, particularly the People's Republic of China and the Democratic People's Republic of Korea, are already talking excitedly about the revival of Japanese "militarism." A change of this magnitude would certainly deepen the confrontation between them and Japan. It would also cause fear throughout the non-communist parts of Asia, for they too were occupied by the Japanese forces only twenty-five years ago. Memories are still too strong for any states in Asia to want to see Japanese troops again on their soil or any large Japanese armada in their waters or skies. These sentiments are changing and time will surely erode these fears if they prove unjustified, but it is still too early to imagine that large Japanese forces deployed throughout Asia or with the capacity for such deployment would do other than to frighten all Asians and would, therefore, add to rather than lessen the instability in the region.

As in the case of the pressures for substantially greater attention to welfare, so in this case of demands for greater concern for prestige and self-reliance in defense, there is no obvious reason why the economists cannot contain them by a dual policy of making judicious but limited concessions while continuing to concentrate fundamentally on increasing the size of the pie from which the shares are divided; but it is very likely that the appeal of economism will decline as the years go by. The reasons are many, but four are especially important. As the economy continues to improve, the willingness of the Japanese to sacrifice for the future will probably decline and the desire to spend in the present, either for private consumption, welfare, or defense, will grow. Militarily, as the defense establishment reaches a size that begins to be felt by the neighboring countries, as it will even under economist policy by 1975, the next step to major-power status will not seem so large. Technologically, Japan will be far closer to relative independence in the aircraft, electronic, space, and nuclear industries. Psychologically, the self-confidence these various developments will breed will inevitably seek a stronger outlet.

Politically also the prospects for a shift in political leadership or a shift in policy by the present leadership in order to keep its position would appear to be stronger in the late 1970's than at present. By then the current moves for a reorganization of the labor movement may well have succeeded, with the result that the socialist parties that depend on labor may have reconstituted themselves. The moderate JSP factions may have joined with the DSP factions in a new centrist party. It is possible that the Komeito also may see the wisdom of joining the new coalition, so that, with the defection of one or more dissident factions of the LDP, the new grouping might then come to power. Of course, it is also possible that the LDP will master its problem of appealing to the new urbanites and will itself continue to rule, but this too will mean new leadership. The LDP faction heads are relatively old men. By 1975 the majority will have left the political scene, clearing the way for a new generation of conservative leaders. Consequently, the welfarist and particularly the chauvinist options can be expected to grow in appeal as the decade wears on.

There are those who have advocated that now is the time for the United States to urge Japan to abandon its policy of balanced defense for one of "burden-sharing." The question posed is: as Japan grows economically stronger, why should it not shift its priorities to give first place to the maintenance of international security in the entire Asian-Pacific region? As the United States finds itself over-extended and seeks to be more selective in its commitments and less forward in its deployments, why should not the Japanese be urged to rearm at a much more rapid pace, take over the commitments and the deployments that the United States would like to give up or reduce, particularly in Korea, the Pacific sea lanes, and Southeast Asia, and thus in general contribute more to maintaining a favorable overall balance of power that the Japanese government finds to be in its interest as well as ours?

There are a number of reasons why this is not a desirable alternative to present policy. First of all, for Japan to be capable of such action, it would need to double or triple the pace at which it is rearming and add a very substantial component of long-range weapons and transport facilities as well as additional manpower with a different kind of training for its forces. From the budgetary point of view it is conceivable that the economy could take this reallocation of resources without hurting its growth rate very much. From the political point of view, however, the Liberal Democratic party could hardly expect to be able to weather the opposition unless the threats to Japanese security were demonstrably worse and the American capacity and will for deterrence were obviously unreliable. Such a situation might occur, for example, if the Russians and the Chinese were able to heal their breach and were to turn a hostile united front toward Japan, or if the Korean war broke out again, or if the United States were to appear to be "bugging out" of Asia, heading for an even larger war in Asia or turning to China in preference to Japan. Even if such circumstances did arise, the government could hardly take up such a policy line without seeking the open support of elements on the right and engaging in a vast propaganda and organizational effort to remold and mobilize the general public.

Secondly, were Japan to rearm heavily under such circumstances, it is not likely that it would do so in a spirit of "what can we do to be helpful to the Americans?" It would rather be in an atmosphere of heightened chauvinism, which suggests not so much sharing America's burdens as going it alone.

Of the realistic alternatives, we must recognize that economism and balanced defense, modified perhaps to give greater attention to welfare, are the policies that accord best with Japanese security interests as well as with our own and those of the rest of the world. It behooves the United States, therefore, in view of the possibilities for change in Japan as well as the probability that American influence in Japan is apt to decline as the decade wears on, to do all it can as soon as it can to strengthen the appeal of the economists to the Japanese people by helping them to achieve those objectives which we share.

Concretely, we must show far greater understanding than we have in the past that the Japanese know what they are doing in adhering to their graduated military build-up rather than pushing to become suddenly the major or even super-military power they are capable of becoming. We should not urge them to quicken the pace, but should support the attainment of their very reasonable objective to become a middle military power by 1975, able to provide the conventional defense of their own islands with their own hands, but not to assert themselves in foreign lands or beyond the adjacent skies and seas. Such support requires a well-coordinated set of military, technological, political, and economic policies.

Militarily, we must continue systematically to reduce our own forces in Asia, returning Okinawa without question in 1972 and looking forward to the evacuation of all or nearly all American military units in Japan by 1975. Our presence will be neither needed nor wanted by that time. This does not mean that we should be working toward a revision or abrogation of the security treaty—just the opposite. It is the security treaty with its promise of support for Japan in any prolonged engagement or threatened nuclear attack that enables Japan to keep its balanced defense policy. It is important, therefore, that, as we complete our withdrawal, we enter into clear understandings with Japan about storage, stand-by, joint, and emergency use of facilities should our return be needed, and that we institute mechanisms for such joint maneuvers and strategic consultation as will make our mutual commitments under the treaty both effective and credible. It is high time, for example, that the annual strategic conferences that have been held for some years on a technical level should be raised to the same full ministerial level as are our annual economic conferences.

Technologically, we must do more than simply offer our advanced weaponry for sale and our conventional military know-how for hire. We must see to it that Japan's decision not to go nuclear does not result in any appreciable penalty for the peaceful uses of the most advanced scientific knowledge. We should, for example, invite the Japanese to share fully in our outer space program, working out arrangements for their scientists to work with ours in preparing for, launching, and evaluating these probes and offering to include Japanese in our astronaut training program and in our crews. We should also give serious thought to the possibility of sharing our nuclear enrichment technology. So long as Japan does not have such technology, it will remain dependent on the outside world, principally the United States, for nuclear fuel; but it has begun serious research in this field. It will certainly find a solution eventually, if not under civilian auspices then finally under the military. Would it not be better to work this out cooperatively rather than competitively? We should consider now whether we could share our technology, perhaps to build an enrichment facility in Japan that would have at least a bi-national character and that might be structured on a multi-national Asian regional basis. Such a regional enrichment facility in which Japan and the United States played leading parts would encourage Japan to keep its nuclear efforts civilian and at the same time strengthen the stability and cohesion of the region. It would also provide an added inducement for Japan to ratify the Nuclear Non-Proliferation Treaty.

We must also see to it that Japan's military self-restraint does not cause the world to underestimate Japan's military potentiality or undervalue its capacity for constructive leadership in world peace. Its desire for a permanent seat on the Security Council of the United Nations should be supported by the United States with enthusiasm.

But these efforts alone cannot succeed in encouraging Japan to adhere to its present security course unless the confrontations of the environment are toned down and a broad spirit of détente begins to pervade the area. While careful not to concede any of its national interests, the government of Japan has clearly been interested in reducing the tensions between itself and the communist states in Asia. The recognition of Outer Mongolia is on its agenda. Relations with the People's Republic of China are constantly being reassessed and every effort is made to keep at least minimal trade with the Democratic People's Republic of Korea and the Democratic Republic of Vietnam to the extent compatible with good relations with the United States. It would like to go farther if it can without permanently damaging its relations with the Republic of Korea and the Republic of China on Taiwan.

This stance would seem to accord rather well with the implications of the Guam Doctrine, for if our friends in Asia are not to be frightened by our action into excessive rearmament and our potential opponents not to be encouraged to try a resort to arms, it will be essential to lower the level of confrontation that now divides them and embroils us.

The experience of the secret Kissinger visit of 1971 shows that for either Japan or the United States to pursue a détente policy alone without full understanding would cause a great concern in the other and would be one of the surest ways to heighten tensions between us. The fact that we now share to a very considerable extent the same conception of the need for détente, if that is possible, as well as deterrence makes this an extremely propitious time to try to devise a joint strategy or simply to coordinate our separate efforts to explore the possibilities for lowering the level of military confrontation with the communist countries.

The problems involved are so complex and so encrusted with history that one hesitates to suggest what lines might be the most fruitful, but among those worth serious exploration would seem to be demilitarization and reassociation in the Korean peninsula.

The United States, for example, might take the lead in exploring the possibility of detaching the United Nations from the truce settlement, possibly dissolving the United Nations Command and the United Nations Commission for the Unification and Rehabilitation of Korea, and making it possible for the Democratic People's Republic of Korea to participate in the UN without conditions. By so doing, the UN might be relieved of a debilitating involvement and restored to a position where it could be a valuable forum for the two Korean regimes to meet. At the same time, with the full understanding of Japan as well as the Republic of Korea, the United

States might continue to reduce its military forces in Korea, looking toward a total or nearly total evacuation by 1975. If accompanied by suitable assistance to the ROK to modernize its own equipment and strengthen its air force so as not to invite attack by the north or instability in the south, such an American withdrawal from its forward position in the Korean peninsula, with the command of Korean forces being turned over to the ROK government, would put the two Korean regimes in a much greater position of military parity and should be a powerful inducement to both to come to an agreement to reduce the military confrontation.

Various questions may be raised. Would Korean replacements for American fighting men on the ground or in the air be as effective as the Americans? In view of the combat experience many Korean fighting men have been receiving in Indochina, it would seem reasonable to assume that they would be. Would the South Koreans, freed of American controls, be inclined to provoke a resumption of the war? There is no reason to assume this, providing that the shift in arms and responsibility is made deliberately, with full consultation and full publicity, so that it is made demonstrably clear to the south that it is indeed strong enough to deter an attack but not strong enough to launch one. But would the north believe this or would it push south again, as it did in 1950 when the American presence was removed? When the American forces were withdrawn in 1947, the peninsular balance of power was drastically upset and the impression was given that we had no intention of ever returning. The proposal here is very different: it is to withdraw only to the extent that parity can be achieved by indigenous forces alone. Moreover, it would seem important in conjunction with any withdrawal that the United States reaffirm its commitment to the security treaty with South Korea and arrange for such stand-by or joint facilities in the ROK and such supply facilities across the Pacific as would permit its rapid return were South Korea to require it, as, for example, if a northern attack were supported by non-peninsular forces. Were these arrangements clear to both sides, it would seem reasonable to believe that both regimes might feel more ready than at present to lower their level of militancy.

Japan, which has some limited access to the north, might be encouraged to explore the possibility of opening up a wider relationship with the DPRK and through such expanded contacts to urge the DPRK and its allies to move toward a peaceful settlement.

Beyond this, it is apparent that any reduction of military confrontation that leads to some form of reassociation of the two halves of Korea will need the approval or acquiescence of each of the great powers whose security is concerned: the USSR, the PRC, and Japan, as well as the United States. Exploring first with Japan and then, if the response is favorable, jointly with the PRC and the USSR, the United States might investigate the possibility of reaching some kind of four-power agreement on the neutralization of the peninsula.

There are many problems involved in these suggestions. By stating them here somewhat baldly, one means only to suggest some of the lines that might be pursued. An intensive study of the possibilities should be inaugurated promptly and exploratory talks with the Japanese begun. Were we able to work out an easing of the Korean confrontation mutually accepted by all the parties concerned, we could probably make no greater contribution to demilitarization of the atmosphere in Northeast Asia.

Central to any effective détente policy should be an effort by the United States and Japan to reassociate with mainland China. It is important that, while our ideas concerning mainland China are so similar to those of the Japanese—China is not a reasonable alternative partner for either of us—we go down the road together to see whether it is possible to get into a better relationship with it. We should be agreeable to

Japan's making whatever efforts it wishes, both politically and economically, asking only that we be kept completely informed. And in turn, we also should continue to try to open the door, not behind Japan's back, but in full consultation with it. We each now acknowledge a relationship with the PRC that amounts to de facto recognition. There is no reason not to explore de jure recognition, leaving it to the PRC to decide whether it wishes to accept this or to impose conditions concerning Taiwan that would be unacceptable.

There is one special concern with China, shared by Japan and the United States but approached from different positions: that of nuclear weapons. It would clearly be in the interest of both Japan and the United States to secure a reliable pledge from the PRC to limit the production and use of such weapons. It would also be in the interests of all other powers. It would seem useful therefore to exhaust every possibility to secure such a pledge. Providing the Japanese are kept fully informed and pledges are not made to China weakening our nuclear umbrella over Japan, there is no reason to fear Japanese hostility to such negotiating, particularly since Japan, in its statement at the time of the signing of the Non-Proliferation Treaty, explicitly called on the nuclear powers to conduct such negotiations.

We might also explore with Japan the mutual desirability of an international agreement on limiting conventional arms supply. Since the reduction of the United States military presence in Asia proceeds concomitantly with the increase in Japan's capacity to produce arms, it is reasonable to assume that in the years ahead more and more of the non-communist developing Asian states will be turning to Japan as a potential arms supplier. So far, the government of Japan has shown great hesitation in this field just as have most leaders of the business community, but these reservations may gradually weaken. It would seem important, therefore, before Japan's attitude changes, for the United States to initiate discussions with it looking toward the conclusion of some kind of agreement on limiting conventional arms sales to the developing countries of Asia.

One specific suggestion might be to see whether the Japanese would be interested in signing a bilateral agreement with the United States incorporating limitations along the lines provided in our own 1968 Foreign Assistance Act and the new Foreign Military Sales Act.[3] One such limitation might be phrased as follows: that each of the contracting parties (Japan and the United States) agree not to supply nor to permit their nationals to supply to the developing countries of Asia sophisticated weapons systems, such as missile systems and jet aircraft for military purposes, unless the president or premier, as appropriate, determines that the furnishing of such weapons systems is important to the national security of his country. Another limitation might be phrased: that when the president or premier, as may be appropriate, determines that a weapons-receiving country is using such weapons for aggressive purposes, or that an aid-receiving country is diverting assistance furnished for non-military purposes to military expenditures or is diverting its own resources to unnecessary expenditures to a degree that materially affects effectiveness of the non-military aid being given, such weapons supply or economic aid will be terminated until the president or premier, as appropriate, is assured that such diversion will no longer take place.

If something along these or similar lines could be worked out, other potential arms suppliers also could be invited to join.

Our economic policy too needs to be formulated with its impact on Japan's security policy clearly in mind. As already pointed out, it is simply unrealistic to think that Japan can

[3] For a fuller discussion of these provisions, see Geoffrey Kemp, "The Availability of Major Weapons Systems as a Constraint on Alternative Indian Defense Programs" (private ms).

play a helpful, direct military role in the Asian balance of power within the next five years. If it continues to arm in the manner and at the pace assumed most likely here, it will still be allocating only half as much money to defense in 1975 as the mainland Chinese or one of the major European powers; and its weaponry, while modern, will be of short range and not deployable on any scale overseas. It will not, therefore, be in a position to use its own forces to assist a friend or deter a foe far from its shores. Nor is it likely to want to. The most it is likely to offer its friends is, toward the end of the period, to supply arms probably commercially and possibly with government aid, and to give technological training programs with military relevance. But it cannot be expected to want to join any military security arrangements obligating it to come to the assistance of others.

On the other hand, Japan can and wants to play a non-military regional role, particularly in providing technological and financial help to the developing nations. Such help would contribute greatly at any time to the stability of these countries since they are all capital- and technology-poor, but it would be particularly helpful in the early 1970's in view of America's changing role and interests in Asia as symbolized by President Nixon's statement when he visited Guam in 1969. In contrast to the previous American posture of containment enforced by a forward positioning of the American military forces and an apparent willingness to engage the communists anywhere, the new posture would seem to be one of deterrence, to be achieved in the military field at least not so much by American forces in the field as by the indigenous forces of these developing non-communist Asian nations themselves.

One consequence of the withdrawal of American forces from forward positions in Asia and the reduction in our general purpose forces at home no doubt will be to stimulate an increasing search by each of the developing, non-communist states in Asia, not so much for new allies—no additions or replacements appear now in sight—as for ways to augment its own military establishment. The demand for arms, military-industrial technology and facilities, and military advice can be expected to rise. Such a demand, if met by a diversion of otherwise productive domestic resources without external assistance, could set back progress toward economic development and political stability in such countries, thereby increasing the likelihood of war.

Two ways to offset such an effect have been suggested already: a broad policy of diplomatic détente and an international agreement on arms supply. Another is to encourage Japan to extend greater economic and technological aid to these countries. The increased extension of such aid could serve to replace such domestic resources as may be diverted to local arms build-ups, thus supporting a healthier development than might otherwise be possible.

The United States should do all it can to encourage Japan to increase the volume of its aid, not only to attain by 1975 the level of 1 percent of GNP it has set itself but also to increase within that amount the share of concessional aid to be given by the government for broad public purposes. We should also urge Japan to take other steps that will relieve its overly favorable trade imbalances with these countries.

At the same time, we must recognize that extraordinarily large aid flows from Japan can also disrupt these countries if the "strings" become too strong. It would be in the interests of all to encourage Japan therefore to transmit as much of its aid as possible through multilateral agencies. Since the United States also is tending to shift the major part of its aid to such instrumentalities, we are in a particularly strong position to take the initiative to urge Japan and other aid-giving countries to do the same.

It is worth emphasizing that for Japan to be persuaded to move constructively along the lines suggested above, it will need to be assured that the United States is moving along similar lines and that Japan's entrance into Asia does not mean America's exit.

But we must do more than this and think more seriously about our own economic policy and its impact on Japan. There has been too much complacency on both sides of the Pacific that the volume of bilateral trade is so great that it forms a bridge of mutual interest that no amount of commercial carping or monetary manipulation can weaken. It should be remembered, however, that the U.S.-Japan trade was great also in the decade before the Pacific war, but volume alone did not suffice to reassure the Japanese about the security of their commercial future.

U.S.-Japan relations have now entered a period of extreme delicacy. Intolerable trade imbalances caused the Nixon administration to seek major changes in the international monetary system and to impose a temporary surcharge that threw up a high wall against Japan's traders. It is to the credit of Japan's leaders that they did not panic, recognizing that in this instance, as in the instance of Mr. Kissinger's secret visit to Peking, the United States was faced with very real problems and the president was acting decisively to try to solve them.

But style in politics is often as important as substance. In all our relations with Japan, and particularly in matters that touch Japan's interests as vitally as our attitude toward China and our international trade and monetary policy, we must learn to be as sensitive to Japan's concerns as we are to our own. Full and frank consultation must be continuous. Can anything less be expected among friends?

CONTRIBUTORS

DONALD C. HELLMANN, Associate Professor of Political Science and Associate Director of the Institute of Comparative and Foreign Area Studies of the University of Washington, Seattle, is the author of *Japanese Domestic Politics and Foreign Policy: The Peace Agreement with the Soviet Union* (1969) and other works. He carried on research in Japan in 1970-1971 as an International Affairs Fellow of the Council on Foreign Relations on the prospects for American-Japanese relations.

FRANK C. LANGDON is Professor of Political Science at the University of British Columbia. The author of the widely used text, *Politics in Japan* (1967), and other works, he has returned recently from Japan, where he was carrying on research for his forthcoming book on Japan's postwar diplomacy.

JAMES WILLIAM MORLEY, Professor of Government and Director of the East Asian Institute, Columbia University, is the author of numerous works on Asia, and most recently editor of *Dilemmas of Growth in Prewar Japan* (1971) and *Japan's Foreign Policy, 1868-1941: A Research Guide* (1972). He served from 1967 to 1969 as Special Assistant to the U.S. Ambassador in Tokyo.

NATHANIEL B. THAYER, Visiting Associate Professor of Political Science at Hunter College, served in the U.S. Foreign Service from 1960-1968, holding among other posts those of Public Affairs Officer in Washington for Japan, Korea, and Okinawa, Press Attaché in Tokyo, and Political Officer in Rangoon. He is the author of *How the Conservatives Rule Japan* (1969).

MARTIN E. WEINSTEIN, Associate Professor of Political Science, the University of Illinois at Urbana-Champaign, is the author of *Japan's Postwar Defense Policy, 1947-1968* (1971).

KENNETH T. YOUNG, formerly Ambassador to Thailand, is one of our most experienced diplomats on Southeast Asia. The author of *Negotiating with the Chinese Communists: The United States Experience, 1953-1967* (1968) and other books, he is currently Senior Visiting Fellow at the Council on Foreign Relations, carrying on a study of America's Asian policy.

Index